BUS

3/9/01

ALLEN COUNTY PUBLIC LIBRAR

03856 6003

P9-EEJ-071

THE
KILLING
ZONE

THE KILLING ZONE

How and Why Pilots Die

PAUL A. CRAIG

McGraw-Hill

New York   San Francisco   Washington, D.C.   Auckland   Bogatá
Caracas   Lisbon   London   Madrid   Mexico City   Milan
Montreal   New Delhi   San Juan   Singapore
Sidney   Tokyo   Toronto

Library of Congress Cataloging-in-Publication Data

Craig, Paul A.
    The killing zone : how and why pilots die / Paul A. Craig.
      p.   cm.
    Includes index.
    ISBN 0-07-136269-X
    1.

  00000.00000 2001
  000.000'00'000—dc21   00-00000

McGraw-Hill

A Division of The **McGraw·Hill** Companies

Copyright © 2001 by The McGraw-Hill Companies. All rights reserved. Printed in the United States of America. Except as permitted under the United States Copyright Act of 1976, no part of this publication may be reproduced or distributed in any form or by any means, or stored in a data base or retrieval system, without the prior written permission of the publisher.

1 2 3 4 5 6 7 8 9 0 DOC/DOC 9 0 9 8 7 6 5 4 3 2 1 0 9

ISBN 0-07-136269-X

The sponsoring editor for this book was Shelley Carr, the editing supervisor was Sally Glover, and the production supervisor was Pamela Pelton. It was set in Janson text per the AV3 design by Pat Caruso of McGraw-Hill's Professional Book Group composition unit, Hightstown, N.J.

Printed and bound by R. R. Donnelley & Sons Company.

This book is printed on recycled, acid-free paper containing a minimum of 50% recycled de-inked fiber.

McGraw-Hill books are available at special quantity discounts to use as premiums and sales promotions, or for use in corporate training programs. For more information, please write to the Director of Special Sales, Professional Publishing, McGraw-Hill, Two Penn Plaza, New York, NY 10121-2298. Or contact your local bookstore.

Information contained in this work has been obtained by The McGraw-Hill Companies, Inc. ("McGraw-Hill") from sources believed to be reliable. However, neither McGraw-Hill nor its authors guarantee the accuracy or completeness of any information published herein, and neither McGraw-Hill nor its authors shall be responsible for any errors, omissions, or damages arising out of use of this information. This work is published with the understanding that McGraw-Hill and its authors are supplying information but are not attempting to render engineering or other professional services. If such services are required, the assistance of an appropriate professional should be sought.

We cannot live long enough to learn from every mistake. That is why it is so important that we get together and share our experiences so that we can learn, understand, and gain airmanship. Together we can attack, and hopefully someday eliminate, the Killing Zone. For now, this book is for the pilots who are today, *flying through the zone...*

P. Craig
Franklin, Tennessee
July 2000

Contents

Acknowledgments

Our flight instructors are the first line of defense in the battle with the Killing Zone. No position in our industry is so important yet so underappreciated. The airline job market is exploding, and most spend only a short time working as an instructor. Nevertheless, the pilots who will fly you and your family in the future are being trained today by a local flight instructor. I have had the privilege to train some of the best in the last year: Josh Fowler, Matt Wolvington, Paul Warren, Nat Harris, Rob Bellamy, Ryan Morford, Adam Byerly, Frank Deal, Victoria Cassel, Paul Mosey, Tony Dobson, Carla McMillen, Don McKelvey, Mitch Lowman, John Penland, Ed Stokes, Chris Smith, Corey Niter, Shane Mowery, Josh Hooper, Chris Childs, Rob Forness, David Moon, Brian Zuelsdorf, James Sperance, and Rebecca Carter. These instructors are male, female, black, white, and Hispanic. They are young, eager, and will soon fill the cockpits of our nation's airliners, fighters, and corporate aircraft. They have all survived the Killing Zone and now help others to follow in their contrails.

Introduction

This book is about aircraft accidents, particularly those accidents involving low-time, inexperienced pilots. It was a hard book to write because it involved fellow pilots in trouble. During the time it took to write this book, two of my pilot friends were involved in accidents. One escaped unhurt; the other was killed. The accidents used for example throughout the book are used only with the greatest respect for the victims. When an accident is evaluated in the book, it is done so in the hope that other pilots might be safer from the knowledge. Many of the accidents used as examples were fatal accidents. By using the fatal accidents of fellow pilots as a learning tool, it is as if they have signed an "organ donor card." Their tragedy may be of value by saving the life of another pilot.

The accidents that are cited here were drawn from the National Transportation Safety Board's public investigation files. The accidents were selected at random. Each was selected for its educational value alone. Accidents from all areas of the country were selected—any disproportionate number in any one area of the country is coincidental. Accidents from different aircraft manufacturers were used as illustration. No attempt was made to single out any manufacturer, and any disproportionate number from any one aircraft manufacturer is coincidental.

CHAPTER ONE

The Killing Zone

DESPITE THE BOOK TITLE, flying is safe. But of course it all depends on your definition of safety. If safety means that an accident can never happen, then flying is not safe. But measured by that standard, nothing that humans do is safe. There is some level of risk that must be assumed in anything. In 1972 the U.S. Supreme Court ruled that, "safe is not the equivalent of risk free." This means that there must be situations in life that carry a certain level of risk but at the same time are considered safe. When accidents happen and people are hurt or killed, even then there is a tolerance level, or an acceptance level at which we still consider the actions safe.

When an airplane accident takes place, the media is quick to find comparisons so that the average viewer or reader can relate the risk involved with the airplane to something they are more familiar with. Sometimes these comparisons take accident statistics out of context and the analogies that are made do not convey the true story.

And of course everything is relative. Did you know that from 1989 until 1993, 1142 private pilots were killed in aircraft accidents. During those exact same years, 228,000 drivers were killed in highway accidents. It would seem that driving is approximately 200 times more dangerous than flying as a private pilot. But these numbers must be compared to the exposure to risk. There were 200 times more fatal car crashes, but there were probably more than 200 times the amount of car drivers than private pilots. The 1999 Nall Report from the Air Safety Foundation reports that motor vehicles have about 10 times as many nonfatal accidents per mile as do general aviation aircraft, but aircraft have about 7 times as many fatal accidents per mile as the motor vehicle.

I am guilty of saying things like, "the most dangerous part of your airplane trip is driving to the airport," and at the conclusion of a flight with passengers saying, "now for the scary part—you must drive home!" But the fact is that when all things are considered within context flying is not safer than driving.

From 1989 to 1993 there were 102 student pilots killed while flying aircraft. We will see throughout this book that student pilots are one of the safest groups of people from any walk of life. During that same time, 4600 died in recreational boating accidents. Boating for fun and flying for fun have often been compared. Boats and airplanes stay docked/tied down at a "port." Both airplane and boat drivers have been called pilots. Most often these pilots take their boats/airplanes out and travel around only to end up right back where they started—at the port.

From 1989 to 1993, 432 passengers were killed in major airline accidents while traveling away from home. During that same time, 105,000 people died due to accidents in the home. During the same time, 132 people were killed on commuter or regional air carriers, sometimes referred to as "puddle-jumpers." Meanwhile 33,500 people were killed in pedestrian accidents (some presumably actually jumping puddles). Fifty flight crew members were killed while on the job between 1989 and 1993. At the same time another 47,000 Americans died on their jobs.

You know the rather dark saying: "Doctors bury their mistakes while pilots are buried with them." A Harvard University study estimates that there are 98,000 deaths each year due to "medical errors." That means, depending on your definition of medical error and what numbers used for comparison, your doctor is approximately 900 times more deadly than your pilot.

You are 100 times more likely to be murdered than to die in a general aviation aircraft accident, and NASA reports that the risk of being struck by a meteorite is comparable to the risk of being fatally injured in an airline accident.

There were 117 deaths due to midair collisions in aircraft and 62,000 deaths from falls that took place not in an aircraft. There were 21,000 deaths from fire, 23,000 people drowned, 26,000 were poisoned, then there was 4000 deaths while bicycling, 3000 deaths at rail-highway crossings, and still another 400 killed in "animal-drawn" vehicles such as wagons and sleighs. What does it all mean? It means that if we live, we must assume risk.

The aircraft accident statistics show smaller accident numbers, but that most often reflects the fact that there are fewer airplanes and fewer pilots than other transportation forms. Are accident statistics even the best way to gauge relative safety? I say no. Just because an accident does not happen, that does not automatically mean that a system is safe. Likewise, the occurrence of an accident does not prove that a system is unsafe. The actual safety of flying is more complex than mere statistics can ever describe. But getting a representation of the reality of safety is tough, maybe impossible. When a pilot does an unsafe thing, but he or she gets away with it and no accident occurs, they are still unsafe. In fact you could argue that getting away with it once without consequence makes them even more unsafe because they can develop an unsafe "attitude," which in turn could lead to more unsafe acts. The accident that does not happen cannot be recorded. It cannot be categorized by statisticians, but an unsafe situation still exists beyond what the numbers describe. The numbers therefore should be considered the tip of the iceberg. The numbers reflect unsafe acts that we can see, but the true level of safety must also include unsafe acts that we did not see.

The accident numbers are not worthless. Even if they do not reflect the complete problem, they do provide insight. The statistics do indicate trends. It is the accident numbers that have revealed that segments of the flying population are more likely to be involved in an accident than others. When compared on a level playing field, it is the general aviation pilot who has the greatest risk among all other pilots of being involved in an accident.

Accident statistics can be evaluated many different ways. You could compare the number of airplanes to the number of airplanes involved in accidents. But that would not account for airplanes that never flew at all to those that are in the air all the time. You could compare passenger miles versus people hurt in an accident. But that would not allow for the fact that some airplanes carry more people and travel greater distances while others carry one passenger and hardly ever leave the traffic pattern. In order to level the playing field statistically, the National Transportation Safety Board (NTSB) relies on a formula that compares accidents that take place for every 100,000 flight hours. This is the best way they have found to compare accidents across the wide range of aviation activities.

The Code of Federal Regulations Part 830 defines an accident as *"an occurrence associated with the operation of an aircraft that takes place between the time any person boards the aircraft with the intention of flight and all such persons have disembarked, and in which any person suffers death or serious injury, or in which the aircraft receives substantial damage."* Figure 1.1 is a table of accidents. It shows that in 1998 there were 48 large air carrier accidents. This yielded a 0.29 accident per 100,000 flight hour rate. That same year, general aviation had 1910 accidents. Now 1910 is a far greater number than 48, but general aviation flies far more hours than the air carriers. Our general aviation aircraft fleet flew an estimated 26.8 million flight hours in 1998. This calculates out to be 7.12 accidents for every 100,000 general aviation hours flown. Figure 1.1 compares large air carriers, commuters, on-demand air taxi, and general aviation from 1995 through 1998. Take note that general aviation has a greater accident rate than the other three categories combined!

The good news is that the accident rate for general aviation is improving steadily. There were approximately 14 accidents for every 100,000 GA flight hours in 1973. The rate dipped below 8 in 1990, but increased in the mid-1990s. Figure 1.1 shows a gradual decrease from 8.23 in 1995 but then a plateau: 7.66 in 1996, 7.29 in 1997, and 7.12 in 1998.

Remember that these are accident "rates." The number of actual accidents can go up and at the same time the accident rate can go down.

Type of Operation	1995		1996		1997		1998	
	Acc	Rate	Acc	Rate	Acc	Rate	Acc	Rate
Large Air Carrier	36	.27	38	.28	49	.31	48	.29
Commuter	12	.46	11	.40	17	1.70	8	1.60
Air Taxi	75	4.39	90	4.44	82	3.64	79	3.11
General Aviation	2053	8.23	1907	7.66	1858	7.29	1910	7.12

Acc = number of accidents   Rate = accidents per 100,000 flight hours

Source: FAA document AAI-200

Fig. 1.1

This is what happened between 1997 and 1998. Accidents increased 3% from 1858 to 1910, but since the total flying hours also increased, the accident rate fell 2% from 7.29 to 7.12. Now here is the scary part. The FAA estimates that general aviation flight hours will increase 9% every four years (source: FAA document: APO-110). That means that this rate could either stay the same or actually improve while the total number of accidents exceeds 2075 accidents in 2002 and 2250 by the year 2006. At that rate, general aviation would be having over 2500 accidents per year by 2010 and, in my opinion that is just unacceptable.

Actually, even one accident is unacceptable. We must reverse the trend. We must drive down the rates by preventing the very next accident. My living is made in aviation. My best friends are fellow pilots. I would hate to think that we did not do everything we could to protect our families and friends from being in one of those 2500 accidents.

General Aviation Flying Is Different

Why do general aviation pilots have more accidents in the first place? Part of the problem is that during those 100,000 hour units, general aviation pilots are exposed to greater risk than air carrier pilots. An airline flight crew has many fewer takeoffs and landings per flight than a GA pilot. They may takeoff once, fly three hours, and land once. On the other hand, GA pilots may fly one hour and do three takeoffs and landings during that hour. As a flight instructor, I know my career take-off and landing versus hours flown rate is much higher than my airline counterparts. Takeoff and landing is more hazardous than straight and level flight. So GA pilots who do more takeoff and landings have great exposure to risk.

GA pilots fly into and out of 18,770 airports (as of 1999). These airports do not all have long, wide, well-lit runways with navigation aids and emergency equipment. Of the 18,770 airports, only 566 were certificated for air carrier operations with aircraft seating more than 30 passengers. In short, our airline pilot friends fly only to airports with the best runways and lots of safety equipment.

General aviation includes aerial application (crop dusting), pipeline and powerline patrol, search and rescue, air ambulance, and many other operations that might involve low-altitude work as a routine part of the job. These GA pilots are certainly exposed to greater risk than an airline captain choosing between the steak or the fish as FL240. The general aviation accident rate is partly higher because every one of our 100,000 hour units contain more hazardous work to be done.

General Aviation Inexperience

But we cannot lay all the blame for the higher accident rate on the type of operations performed. No, the largest contributor to the accident rate in general aviation is inexperience. But it is wrong to think of general aviation pilots as a group with little or no experience. Remember, general aviation is anything that is not a scheduled air carrier or military. This is a wide range. This means that the pilots of transoceanic, multi-million dollar business jets are GA pilots. Air ambulance pilots who save lives by risking their own are highly skilled GA pilots. Even your local airport flight instructor who makes a living passing on the gift of flight is a GA pilot. These people are pros. It is not a fair characterization to label GA pilots as "joy riders" in their "little" airplanes, buzzing and hedge-hopping around on the weekend. Many GA pilots are holders of the Airline Transport Rating even though they never will fly for the airlines.

If we look at inexperience as a cause of accidents, we are not looking at the whole of general aviation, but a smaller subset. Most people do learn to fly inside general aviation and then build their experience as GA pilots. During this experience-building time, the new pilot seems to be caught between two worlds. On the one hand they are fully licensed and legal pilots, but on the other hand they have not had the opportunity yet to learn from experience. Is it true that experience is the best teacher?

When my private pilot checkride was completed, my examiner said, "I'm going to give you your license to learn." I did not know what he meant. I did not know that was his way of telling me that I had passed. But the examiner knew that with my private pilot certificate in hand I would learn much more as a pilot than when I was in training to be a pilot. In my head, I'm thinking that the training is over—after all I did pass! I was so naive. I thought all the learning took place while preparing for the test and that the learning stopped after the test. I was not only inexperienced as a pilot, but I also had an inexperienced "attitude."

As I flew I did learn more, lots more. I scared myself a few times along the way, but I survived my own ignorance. I look back on myself during those years and think how dumb, even dangerous, I was then. Of course at the time I did not think I was dumb or dangerous. I always thought I was a good pilot at the time. This just proves that we always are learning and we can never be stationary or arrogant. I think I'm a pretty smart and safe pilot right now, but I hope I learn more in the next five years so that I can look back on myself today and say, "what a dummy I was back then."

I survived my own ignorance, but others were not so fortunate. Some pilots along the way did not learn from their mistakes, but unfortunately were killed by their mistakes. In a way, the definition of an experienced pilot is a pilot who has *simply survived themselves*.

How big a factor is inexperience? Are accidents spread out among pilots with varying amounts of experience, or are the accidents centered around the inexperienced group? Common sense would say that under-experienced pilots would have the greater amount of accidents, but is there proof of this concept? I wanted to know for sure. I obtained the accident reports from the National Transportation Safety Board (NTSB) that cover the years from 1983 until 2000. I was particularly interested in those years because I first became a flight instructor in 1983. I used the database of accidents and cross-referenced accidents that occurred among pilots during their first 1000 hours of flight experience. The results proved our greatest fears: inexperience kills.

Figure 1.2 is a chart of all fatal accidents during the 17-year period from 1983 to 2000 among private and student pilots of single-engine airplanes. The problem jumps out at you when you first see the chart without even looking at any of the numbers. It is not a "bell" curve. Statisticians would call what this looks like a "skewed distribution." This chart has a total of 2501 fatal accidents depicted (Appendix A1 has a table with the exact numbers for each time block). But of this 2501, the majority of the accidents are on the left side (1874 accidents at 500 hours and less). The left side represents inexperience. In fact, 57% of all the accidents shown on the chart took place when the pilot had between 50 and 350 flight hours. Past 350 hours of flight experience, the accident numbers drop off and then gradually reduce. The "hump" in the chart is significant. Remember these are not just cold numbers but aircraft accidents where people, possibly friends of ours, lost their lives. The chart clearly illustrates a zone where accidents are mostly likely to happen, and since these are fatal accidents this could only be called *the killing zone*.

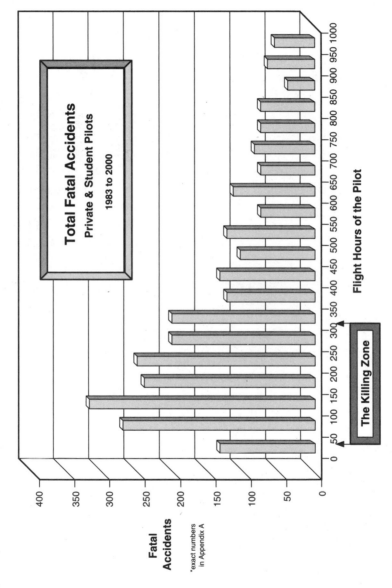

Fig. I.2

8

Attacking the Zone

What has been done in the past to attack the killing zone? In 1986 the FAA made an attack on the zone by changing the time requirement for the instrument rating. Rather than increasing the time needed for the rating, the FAA *reduced* the time needed to qualify for IFR. Before 1986 the regulation required a pilot to have 200 hours of flight time before becoming eligible for the instrument rating checkride. In 1986 that total flight time requirement was dropped to 125. This seems backwards when you first hear it. If more experience is better, why would the FAA allow pilots to fly IFR with less experience? It would seem that this change would place even more underexperienced pilots in harm's way. But it was not the IFR pilots that the FAA was worried about. It was the VFR pilots who were "building time" in order to qualify for the instrument rating.

Most people earned their private pilot certificate with between 50 and 60 flight hours. When the instrument rating was at 200, private pilots had approximately 140 hours to go to make up the gap between private and instrument. The instrument rating itself required 40 hours of training, so in a best-case scenario students would get their private pilot certificates at 50 hours and later begin instrument training at 160 hours (or 40 hours to go before reaching 200). This meant that the pilot had to "bore holes in the sky" for 110 hours to make up the difference! These 110 hours are in the heart of the killing zone. The 110 flight hours were being flown by new and inexperienced pilots, many who wanted to be instrument rated but who were prohibited by the aviation laws of that day from becoming instrument rated.

The regulation change that took place in 1986 was the product of many years of work and study. In 1974 the NTSB published a "special study" about weather-related general aviation accidents. Figure 1.3 is a chart of the study, which covered the years of 1964 through 1972. Even though the number of years represented is less, and only "weather-related" fatal accidents are listed, another killing zone is evident. There were 1910 such accidents during those nine years. Of those, 476 accidents, or 25%, involved pilots with between 100 and 299 flight hours. The NTSB final report made this conclusion: *"Pilots with fewer flight hours were more frequently involved in weather accidents, especially those pilots with more than 100, but less than 300, total flight hours. Perhaps the*

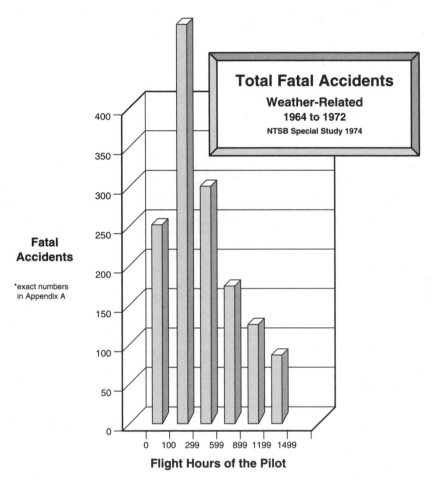

Fig. 1.3

explanation for the peak is that by the time a pilot has accumulated 100 to 299 hours, he is confident of his flying ability even though his actual flying experience is low." The 1974 NTSB report pointed to a pilot's inexperience mixed with a dose of overconfidence as a fatal mix.

The 1974 study ended with a "statistical summary" that made a profile of the pilot who would most likely be involved in a fatal, weather-related, accident. The profile said that the accident pilot:

1. Had received an adequate preflight weather briefing.
2. Was on a pleasure flight.
3. Had less than 100 flight hours in the aircraft being flown.

4. Had a private pilot certificate.
5. Had been practicing for instrument flight with between 1 and 19 hours of simulated instrument time.
6. Had not filed a flight plan.
7. Was accompanied by at least one passenger.
8. And had between 100 and 299 flight hours of experience.

That profile sounds like almost every flight that I took when I had between 100 and 299 flight hours. I was the pilot in the fatal profile!

It was on the evidence presented in the NTSB special study and others that the FAA acted. The strategy was this: If pilots are killing themselves because they are flying to build time, why not eliminate this time-building segment from the equation? Place the instrument rating time requirement right in the middle of the problem. If a person earns the private pilot certificate at an average of 50 hours, and if the instrument rating requires 125, the gap is now only 75 hours instead of 150. The instrument rating still required 40 hours of instrument training, so if a pilot started instrument training at 85 hours so as to reach 125 total at the same time as reaching 40 instrument training, then the gap would only be 35 hours. Instead of 110 hours of "boring holes in the sky," there would only be 35. The idea depended on the fact that instrument training was safer than unsupervised VFR flying.

The plan seems to have had an effect. There were approximately 14 accidents for every 100,000 GA flight hours in 1973. The rate was 7.12 in 1998. Today the minimum time requirement has been eliminated altogether. Now there is no "total time" minimum for the instrument rating. You must meet the 40 hours of instrument instruction requirement and be proficient enough to pass the checkride, regardless of your total time.

Student Pilot Safety

Student pilots are the most inexperienced pilots of all, but they only had 5% of the accidents shown in Figure 1.2. As a group, student pilots, are a very safe bunch, but it should not take a genius to figure this out. Student pilots either fly with a more experienced flight instructor or they fly solo, but only when the flight instructor determines that it is safe. Two heads are better than one, especially when one has been around a little. The flight instructor's role is not only to teach, but also supervise. The flight instructor must endorse a student's logbook for the initial solo flight and for any flight away from the home airport. This means that the flight instructor holds veto power. I have often canceled a student pilot solo flight because, in my opinion, the proper combination of factors were not all present. I look at the weather, wind, time of day, distance of the proposed trip. I try to get a sense of the students' "readiness" for the flight. Are they thinking clearly, do they seem to be in a rush, have they forgotten several items? If any of these factors don't seem quite right, I reschedule the flight. At the same time I don't want my students to expect the decisions to always be made by me—after all, I'm training them to make decisions. So usually, when facing a cross-country flight, I have the student get a weather briefing and prepare a navigation record. Meanwhile, I get a separate weather briefing and make a judgment on the flight. I want my student to come to me with a go/no-go decision. About 90% of the time my decision and the student's decision match. When they don't, I help the student see what factors are determining my decision. In the end the students do not make their own decisions—the instructor has final veto power.

What happens after the student pilot becomes a private pilot? The instructor no longer holds the decision veto. The complete power to decide now is owned by private pilots themselves. Passing the private pilot (or recreational pilot) checkride is like graduation day. Students pass from a protected, regulated environment to a world on their own. Unfortunately the numbers indicate that some people are just not ready to handle this kind of responsibility. From 1983 to 2000 there were 130 fatal accidents among pilots with 49 hours or less—student pilots in a sheltered environment. But the accidents doubled—from 130 to 259—among pilots with 50 to 99 flight hours. Once out from under the authority of the flight instructor, fatal accidents are twice as likely to happen.

Private Pilot Responsibility

Many people say that the reason they learned to fly, or why they want to learn to fly, is for the "freedom." But there is not much freedom when a flight instructor has the complete power to tell you what to do.

Some quickly associate checklists, inspections, and safety precautions with the shackles placed on them by the instructor. For these people, checkride day is liberation day. This is the day when they truly achieve the freedom that they had been looking for all along. Now nobody can tell them what to do. Those "extras" the instructor used to preach about were just overkill that student pilots probably do need in the beginning, but that licensed pilots need not worry about anymore. This attitude leads to accidents and has created the killing zone.

The 1974 NTSB special report alluded to this attitude: *"Perhaps the explanation for the peak is that by the time a pilot has accumulated 100 to 299 hours, he is confident of his flying ability even though his actual flying experience is low."* The killing zone is actually a dangerous overlap. Pilots have been given the responsibility of pilot in command, but at the same time they do not have the maturity needed to respect what it means to be pilot in command. I heard it said another way once about a pilot who had used poor judgment: *"His ego is writing checks that his experience can't cash!"*

Inexperience and Pilot Hiring

Most pilots do not have a dangerously overconfident attitude, but we cannot escape the fact that a lack of experience is inevitable. You must have 10 hours before you can have 20. You must have 100 hours before you have 500. So since a passage through inexperience is inevitable, does that make the killing zone inevitable? It may be, but I'm not willing to let it happen without a fight! I believe the accident numbers can be lowered. I believe the killing zone can be attacked, but it is the age-old catch 22. If pilot "expertise" naturally comes from experience, how can we gain "seasoning" while we are yet inexperienced? How can we fly with the knowledge gained from 1000 flight hours when we only have 100? The answer is to make those first 100 really count. Those first 100 hours, or 300 hours, or 1000 hours cannot be just "boring holes in the sky." To be smarter we must train smarter. The remainder of this book will point out the problems and target smarter solutions. The way to eliminate the Killing Zone is to "train-out" the next accident that would have otherwise happened.

Working to reduce or eliminate the Killing Zone is the goal that all of us in aviation should work toward, but this problem is bigger than just the aviation community. Our national economy depends in many ways on aviation. People are touched by aviation every day, even if they never fly in an airplane.

People who do fly are flying more. Airline companies are buying airplanes and hiring new pilots. In 2000 for the first time, every major airline was hiring pilots. The major airlines get their pilots from the regional and commuter airlines. The regional and commuters get their pilots from cargo haulers and flight schools. The laws of supply and demand seem to work perfectly for pilot hiring. When few pilot jobs are available, the minimum flight time for new hires goes up.

In the early 1990s, the minimum requirement to land a regional air carrier job ranged between 2000 and 2500 hours of flight time. The applicants had to already have the airline transport pilot certificate and at least 500 to 1000 hours of multiengine flight time. When new pilots did get jobs, they very often had to pay for their own training by giving the airline $10,000.

Ten years later things were different. Everybody from top to bottom was hiring. Flights were being canceled because airlines did not have crews to fly them. The magic number to be hired by the regionals dropped to about 1000 hours total time with as little as 50 hours multiengine time. One thousand hours does not qualify a pilot for the ATP so that requirement was dropped, uncorrected vision requirements were dropped, and some regional air carriers even started paying a signing bonus to ensure that they got somebody to fill their flight decks.

The rapid flow of pilots up the career ladder creates some unsettling situations. The biggest problem became the lack of experienced regional airline captains. Since flying began, the time-honored ritual for first officers was to fly with an old-time, crusty captain who would show them the ropes. But with rapid upward pilot hiring, the captain-qualified pilots do not stay with the regionals very long. The upgrade time from first officer to captain was an apprenticeship that lasted 5 to 10 years. Now the upgrade is only 5 to 10 months. This means there are no old-time, crusty pros to learn from. The captain of any given regional flight was probably not even an airline pilot one year ago. It was also common wisdom that at the very least a pilot should be a first officer for one year, if only to experience a year's worth of varied weather conditions. But when pilot expansion is taking place, there is no time for anyone to stay in one place for a full year.

At the beginning of 2000, the U.S. Congress passed a bill that would make it easier for the users of aviation to get money from the National Aviation Trust Fund. The money in the fund comes from the tax we all pay when we purchase an airline ticket, ship a package overnight, or buy a gallon of 100LL aviation fuel. In theory the money should be used to improve airports, upgrade radar and air traffic control, and many other aviation needs. Over the years as flying has increased, the amount of dollars in the trust funds reached into the billions. The problem was that those billions made the deficit and national debt look smaller so Congress and past presidential administrations were stingy with "our" money. The 1999 fight was to take the trust fund "off budget," in other words, take control of the money away from the people who appropriate all other spending. The problem was that control of money equals power in Washington, so many people who had the power did not want to give it up. During 1999 that bill was being debated, argued, and eventually crafted. I made several trips to Washington D.C. that year for meetings related to the bill. My concern was for flight training and research. One particular meeting I had was with the members of the House Aviation Subcommittee and the House Science committee. I told them a story that I hoped would impact their thinking on pilot training, on experience, and indirectly to the Killing Zone. Here was the story:

A former student of mine landed a flight officer's position with a regional air carrier. He was typical of thousands of other young and relatively inexperienced pilots getting hired. After six months on the job he was put into a captain's upgrade class. He was being pushed to upgrade, not because the company believed he was completely ready, but because several captains had departed and they simply needed another captain right away. The captain's transition course was not a separate course. Pilots moving to the captain's seat sat in for a systems refresher with a new-hire class. Then when the new hires went to lunch, the new captains spent an extra hour going through the airline's manual. "They just told us that, as captain, if anything goes wrong it would be our fault!" There was no extra instruction on how to "be" a captain or how to prevent something from going wrong in the first place. Two weeks later the new captain got snowed in at an outlying airport. He had been an airline pilot less that a year. He had never been an airline pilot in the winter. For several days, flights were canceled.

Finally the weather improved and he and his first officer were going to fly back to the hub. When the captain arrived at the airplane, it was still covered in snow and ice. His first officer asked some questions about the deice procedure. That is when it hit him. Not only had he never flown an airplane before that needed to be deiced, he had not even seen it done. He had only read about the procedure. The new captain did not want his first officer to think he did not know what he was doing, so he gave him a few fleeting words and he invited the paying passengers onto the airplane. Soon they were on their way. He told me later, "I guess the deice guys knew what they were doing, because we didn't crash."

The congressional staff members found this story hard to believe. One said, "I'm taking the bus back to my district."

The problems associated with a pilot's lack of experience used to be reserved to general aviation flying. It always seemed that airline flying was safer because a wiser, time-tested veteran was at the controls and this guy could calmly handle any situation. That assurance is now no longer there. The Killing Zone presents a problem for people who are pilots, people who are passengers, and for all people who are a part of our aviation-dependent economy.

CHAPTER TWO

The Dangers

IT SEEMS LIKE a very cold thing to refer to people who have been in accidents as numbers. I try to think of it differently. I try to understand the people and their motivations before the accident. Year after year pilots make the same mistakes. Year after year pilots make the same poor decisions that lead to predictable accidents. I want to know if we can do anything before the fact.

Unfortunately, it is possible to predict with reasonable accuracy how many general aviation pilots will have a fatal accident this year and it is also possible to predict what they will be doing when the accident happens. So if we know in advance what will happen, can we get to these pilots in time? Can we train-out the next accident?

The accident numbers, as described in the first chapter, are all "normalized" by comparing accidents with a standard. That standard is usually accidents per 100,000 flight hours. This levels the playing field and makes yearly comparisons possible. In general aviation, the amount of flight hours in a year is estimated by the FAA. Figure 2.1 illustrates the fluctuating GA flight hours over the past two decades. The total flight hours in general aviation has a high correlation to the national economy as a whole. Flight time in the 1980s was much higher every year than in the 1990s, but the late 1990s showed a steady climb in activity.

Figure 2.2 illustrates the fatal accident rates during the same years. The rate is never lower than one fatal accident per 100,000 hours, or ever higher than 2. Comparing Figure 2.1 with 2-2, you can see that several times the amount of flying went down, and at the same time the accident rate went up (1991–1992 as an example). Does this suggest

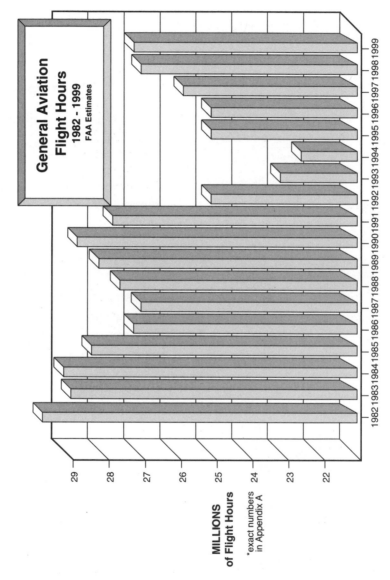

General Aviation
Flight Hours
1982 - 1999
FAA Estimates

1982 1983 1984 1985 1986 1987 1988 1989 1990 1991 1992 1993 1994 1995 1996 1997 1998 1999

29 28 27 26 25 24 23 22

MILLIONS
of Flight Hours

*exact numbers
 in Appendix A

Fig. 2.1

18

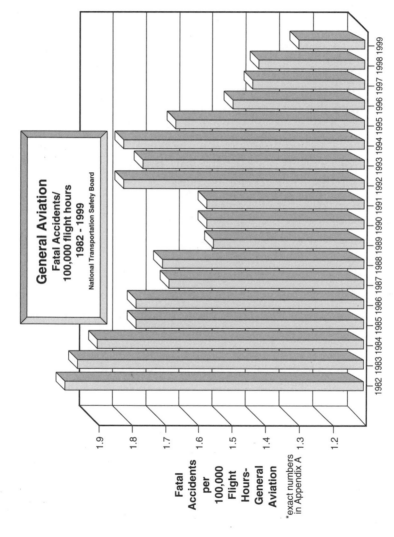

General Aviation
Fatal Accidents/
100,000 flight hours
1982 - 1999
National Transportation Safety Board

Fatal Accidents per 100,000 Flight Hours- General Aviation

* exact numbers in Appendix A

1982 1983 1984 1985 1986 1987 1988 1989 1990 1991 1992 1993 1994 1995 1996 1997 1998 1999

1.9 1.8 1.7 1.6 1.5 1.4 1.3 1.2

Fig. 2.2

that people have more accidents when they fly less due to a lack of proficiency? There are other times when the amount of flight hours increased and at the same time accidents decreased (1996–1997). That must have been a really good year where there were more flights and more happy endings.

Figure 2.3 illustrates the number of total accidents per 100,000 flight hours over the 1982–1999 time frame. This chart includes fatal accidents, but also all accidents where minor injuries and no injuries resulted. Again, when flight hours were at their lowest (1994), accidents were at their greatest in ten years (8.96/100,000). These numbers seem to be telling us that for pilots, flying less is more dangerous than flying more.

The steady decline in accidents in the late 1990s, ending with the lowest rate of accidents since statistics have been kept (7.05/100,000 in 1999) is something the general aviation community should be quite proud of. It also says that an increase in pilot knowledge saves lives and prevents accidents. The sale of aviation safety-related books and magazines went up, and attendance at safety seminars and pilot proficiency programs also increased during these years.

But it is still not enough. Although at an all-time low, accidents still happen and people still get hurt and some die. And it's the same old story when it come to pilots with less experience. Even when the overall rates are low, the proportion of those in accidents with low experience is still high.

Figure 2.4 is a look at the fatal accidents among those pilots who had between 50 and 350 flight hours—the pilots of the Killing Zone. Within this group there is good news. The rates inside the zone have dropped faster than those outside the zone. The five-year average of fatal accidents from 1983 to 1987 was 110 among pilots who had from 50 to 350 flight hours. The five-year average from 1994 to 1998 was 62.5 for the same group. That represented a 40.8% drop in fatal accidents. Comparing the same two time periods among pilots with between 350 and 1000 flight hours, there was a 28% reduction. This means that the Killing Zone was once worse than it is now, but the total number of fatal accidents within the zone are still significantly higher than with any other group.

It is still possible to predict with surprising certainty how many accidents there will be next year, who will be involved in these accidents,

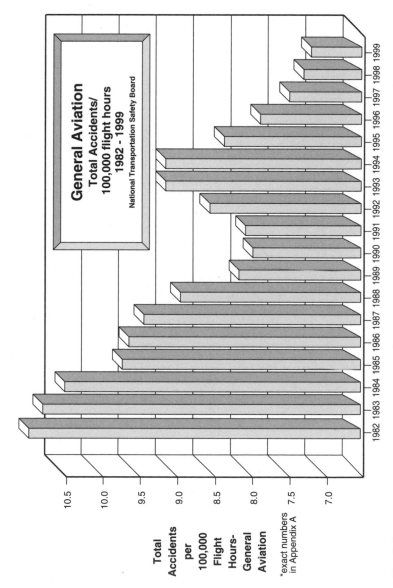

General Aviation
Total Accidents/
100,000 flight hours
1982 - 1999

National Transportation Safety Board

Total
Accidents
per
100,000
Flight
Hours-
General
Aviation

*exact numbers
in Appendix A

Fig. 2.3

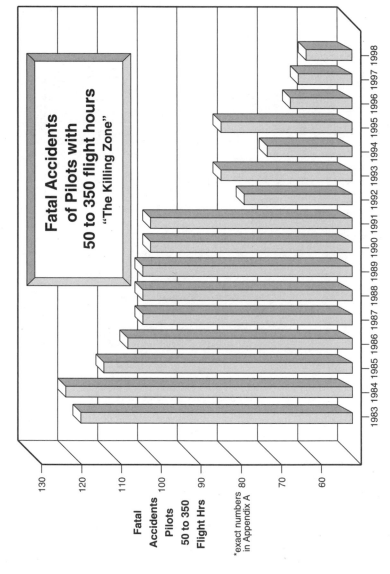

Fig. 2.4

and what they will be doing when the accident occurs. We cannot tell the future, but if the accidents of the past reoccur from year to year, then we can predict the future. If pilots, as a group, make the same mistakes again and again, then looking at the past accidents is almost like looking into the future. What are these mistakes and who is making them?

Accident Investigation

When a general aviation accident occurs, an investigation is conducted. Unfortunately the level of investigation varies widely. It is also true that the level of investigation has much to do with who was in the accident. A celebrity in an accident will draw many more resources to an investigation than when a noncelebrity is involved. We were all happy that John F. Kennedy, Jr. and his family were recovered from the Atlantic, but there are many people who have never been recovered in similar accidents because they were not the son of a president.

Officially the National Transportation Safety Board conducts all accident investigations. The NTSB is a separate government agency that is charged with determining the "probable cause" of accidents. After an investigation is concluded, the NTSB routinely will make recommendations so that similar accidents can be avoided in the future. These recommendations are made to the FAA, aircraft manufacturers, weather observers, or anyone else whom the NTSB believes could have a positive impact. The NTSB is kept separate from the FAA to avoid any conflict of interest. Often the NTSB recommends that the FAA change laws or procedures. But there is a problem. When a "low-profile" accident occurs in general aviation, the NTSB does not often investigate. Instead, they empower the local FAA office to investigate on their behalf. This takes place in approximately 5% of the investigations. When the NTSB delegates to the FAA, any conflict of interest that the FAA might have does not have a check or balance. The FAA does not have a great track record when it comes to investigating itself. This does not necessarily mean that general aviation accidents are not investigated well, but the fact remains that general aviation accident investigation has limited resources.

In Chapter 1, I made the case for being careful with accident statistics. The lack of accidents does not guarantee that flying is safe, and a spree of accidents is not proof that flying is dangerous. Now I reduce the reliability of the numbers further by suggesting that general aviation accident numbers may or may not be the product of full NTSB investigations. The numbers are watered down, but they still point to trends and they point to circumstances where accidents have routinely happened in the past.

The statistics show that there is a list of situations where accidents most likely occur. The list is:

Continued VFR flight into IFR conditions
Maneuvering flight
Takeoff and climb
Approach and landing
Runway incursion
Midair collision
Fuel mismanagement or contamination
Pilot health and physiology
Night flying
Encounters with ice
Instrument flight
Transitioning to advanced aircraft

If accidents from this list were eliminated, there would be very few accidents. But elimination of accidents from these areas has been quite elusive. Despite all efforts up to this point, these categories remain the big 12 pilot killers. And in each category, the lack of experience continues to contribute to accidents. The following chapters examine each killer in detail, in hopes of making a dent in the numbers by substituting knowledge for experience.

CHAPTER THREE

Continued VFR into IFR Conditions

FLYING INTO BAD WEATHER causes the greatest number of aircraft accident deaths. If you compare the total number of weather-related accidents to the number of weather accidents where someone was killed, the fatality rate is high. In 1998, 83% of weather accidents were fatal. Weather accidents usually have three phases, and these three problems combined create the deadly situation. First the pilot flies into bad weather or deteriorating weather. Second, the pilots loses control of the airplane or descends too low in an attempt to fly under the clouds. Third, the airplane strikes an object or the ground at a high rate of speed. So weather creates accidents in many categories. Stall/spin accidents happen when airplane control is lost. Controlled Flight into Terrain (CFIT) accidents happen when pilots try to duck under low clouds. Structural failure accidents happen when disoriented pilots overstress the airplane in an attempt to regain control. Weather accidents are simply an avalanche of events that can sweep a pilot into a situation that cannot be overcome. Once this chain of events starts flowing, there is very little a pilot can do to stop it. The only solution is to prevent the first step. But pilots continue to fly into bad weather even though they, at that point, have control of the situation. Between 70% and 75% of all weather-related fatal accidents start when the pilot, "attempted VFR flight into instrument meteorological conditions (IMC)." The pilot voluntarily flew into the clouds or into an area of poor visibility.

There are many examples of this three-step (continue VFR into IMC, lose airplane control, strike the surface) type accidents. But unfortunately the most famous of these was the accident of John F. Kennedy, Jr. The National Transportation Safety Board's preliminary report (NTSB identification: NYC99MA178) gave this account of this flight from VFR into IMC:

On July 16, 1999, about 2141 eastern daylight time, a Piper PA-32-R301, Saratoga II, N9253N, was destroyed during a collision with water approximately 7.5 miles southeast of Gay Head, Martha's Vineyard, Massachusetts. The certificated private pilot and two passengers were fatally injured. According to computer records, a person using the pilot's subscriber log-in code obtained aviation weather information from an Internet site at 1834. The weather information was for a route briefing from Teterboro, New Jersey to Hyannis, with Martha's Vineyard as an alternate. The forecast for Hyannis called for winds from 230 degrees at 10 knots, visibility 6 miles, and sky clear. All airports along the route reported visual meteorological conditions. The flight departed Essex County Airport at 2038. The pilot informed the tower controller that he would be proceeding north of the Teterboro airport, and then east bound. There is no record of any further communications between the pilot and the air traffic control system. According to the radar data, the airplane passed north of the Teterboro airport, and then continued northeast along the Connecticut coastline at 5,600 feet, before beginning to cross the Rhode Island Sound near Point Judith, Rhode Island. A review of radar data revealed that the airplane began a descent from 5,600 feet about 34 miles from Martha's Vineyard. The airspeed was about 160 knots, and the rate of descent was about 700 feet per minute (fpm). About 2,300 feet, the airplane began a turn to the right and climbed back to 2,600 feet. It remained at 2,600 feet for about one minute while tracking a southeasterly heading. The airplane then started a descent of about 700 fpm and a left turn back to the east. Thirty seconds into the maneuver, the airplane started another turn to the right and entered a rate of descent that exceeded 4,700 fpm. The altitude of the last recorded target was 1,100 feet. On July 20, 1999, about 2240, the airplane was located in 116 feet of water, about ¼ of a mile north of the 1,100 foot radar target position. Preliminary examination of the wreckage

revealed no evidence of an in-flight structural failure or fire. Two of the three landing gear actuators were recovered and found in fully retracted position. There was no evidence of conditions found during examinations that would have prevented either the engine or propeller from operating. Pilots who had flown over the Long Island Sound that evening [evening of the accident] were interviewed after the accident. These pilots reported that the in-flight visibility over the water was significantly reduced.

Just less than a year after the accident the NTSB issued the final report. It listed the probable cause to be: *The pilot's failure to maintain control of the airplane during a descent over water at night, which was a result of spatial disorientation. Factors in the accident were haze, and the dark night.*

Weather accidents have few survivors and therefore there is never much information about what exactly took place. But from the evidence, this accident appears to be quite typical of weather accidents: A pilot flies into IMC, control is lost, the airplane strikes the surface, causing people to die. The weather reports seemed to indicate good weather and good visibility, but over the water, where weather is not reported, things were different. I have a friend who did fly over the Long Island Sound about the time of the accident. He is an airline captain and he was approaching New York from London. He told me that over the water that night it was completely without horizon and ground reference. The investigation tried to eliminate other causes: The engine and propeller were operating, there was no fire, there was no preimpact structural damage. So we know that the plunge to the surface was not deliberate. A pilot with an inflight fire might dive to the surface on purpose in order to get out quick—but this did not happen in this case. The airplane came down out of control even though the airplane was operating properly.

John F. Kennedy, Jr. was a famous man, the son of a president, but as a pilot he was not unusual. He was a private pilot. He was not instrument rated and had just over 300 flight hours experience. Figure 3.1 illustrates fatal accidents among private and student pilots when weather was listed as the accidents' "broad cause." The chart shows again that student pilots are about as safe as higher-time pilots in the 700-flight-hour and above range, but that private pilots with between 50 and 350 hours show a marked increase in accidents. Weather-related accidents have a Killing Zone.

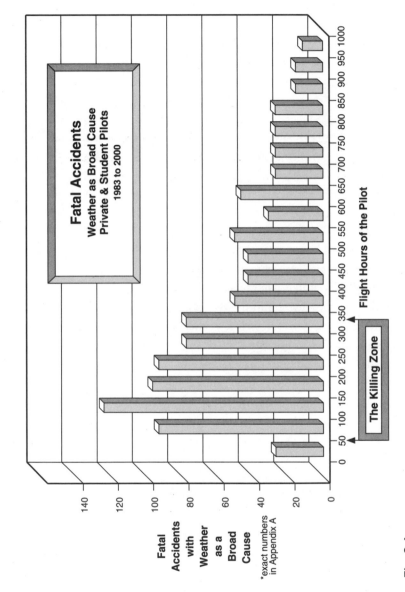

Fig. 3.1

28

Figure 3.2 depicts a more narrow slice of accident data. Figure 3.2 shows the fatal accidents that occurred when VFR pilots attempted to fly into IFR conditions. It is scary, but no longer that surprising: Pilots who have total flight time between 50 and 350 flight hours of experience fly into clouds and kill themselves more often than any other group of pilots. The JFK, Jr. accident is included in Figure 3.2.

Killing Zone Survivor Story

Building cross-country time for my instrument rating, that was my objective. Due to three weeks of weather delays, this flight had been alluding me. But today the weather looked like I could finally make the trip. My plan had been to fly to Memphis from Murfreesboro, Tennessee, with a precautionary fuel stop in Jackson, Tennessee. On the way back this fuel stop would be unnecessary because the winds would be pushing me along.

Feeling pretty confident in my abilities, since I recently updated my logbook and found that I was now closing in on 150 hours, I was ready to tackle Memphis. I went through my preflight planning, which included talking to Flight Service for the current and forecast weather. I filed my VFR flight plan and headed out to the airplane.

The Cessna 152 was topped off and preflighted in no time. Shortly thereafter I was defying gravity once again like an old veteran. My flight to Memphis, including my fuel stop, went off without a hitch. The weather was good. I handled the busy ATC communications very well and even navigated myself to the FBO once on the ground. If I remember right, I even had a squeaker of a landing. I put in my fuel request to be topped off and went in to have a quick snack, pick up some new weather, and file a new flight plan.

How could things be better? I was taxiing out with the big guys and I felt confident. The sun was setting and weather was looking great. I took off with the sunset at my back and headed home. Now that class B airspace was behind me, I concentrated on pilotage. My navigation was going really well, I was finding all of my checkpoints and crossing them right on time. The wind forecast was holding true.

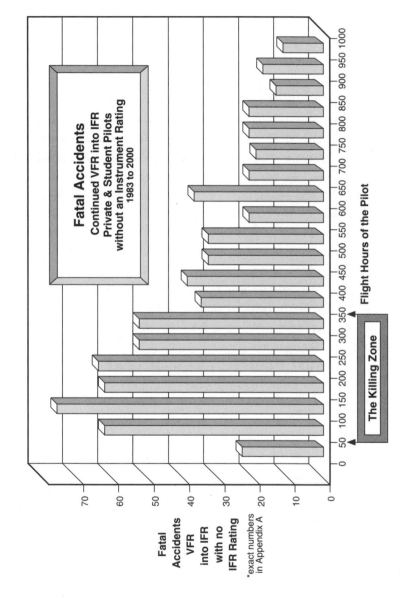

Fatal Accidents
VFR
into IFR
with no
IFR Rating

*exact numbers
in Appendix A

Fatal Accidents
Continued VFR into IFR
Private & Student Pilots
without an Instrument Rating
1983 to 2000

The Killing Zone

Flight Hours of the Pilot

Fig. 3.2

By the time I passed near Jackson, Tennessee, night had set in. I started to notice that it was getting more difficult to pick up the crisp image from the town's lights that were helping me identify my checkpoints. It seemed to be more of a glow now. To ease my troubled mind, I called up Flight Service to get an update on the weather. Flight Service informed me that there wasn't any change in the forecast and that the weather along my route was excellent VFR. This report somewhat set my mind at ease, so I pressed on even though I did not feel 100% comfortable.

As I approached the Tennessee River, nothing had improved. In fact, it had gotten much worse. Before I knew it, I couldn't see anything besides what looked to be directly below me. My comfort level was out the window at this point and I needed some guidance. I made my third call to Flight Service only to hear the same reports. Clear sky and better than 10 miles visibility over the entire route. It just didn't make sense. How could they be reporting such good weather when I can't see anything? Enough was enough! I was the only one that was going to get me out of this so I needed to stay in control and break the chain of events that link together to cause bad things to happen.

I had already tried lower altitudes to maintain ground contact, but I didn't feel comfortable going any lower. Giving up on the idea of trying to navigate by ground reference, I tuned in the Graham VOR and proceeded directly to it in hopes of landing at Centerville, Tennessee.

I started making my radio calls as I approached the airport and I could see the beacon through the haze. Finally, even though the haze was terrible, I had the airport in sight with no more than 4 miles visibility. I overflew the field and entered one of the ugliest patterns I have ever flown. I was doing everything possible to keep the runway in sight while making my calls and configuring to land. I thumped down on the runway with a bounce or two, being thankful, probably for the first time, to be on the ground rather than in the air. I taxied in to a closed FBO, found a phone, and canceled my flight VFR flight plan.

Being upset and embarrassed about everything that had just happened and wondering if I was going to get in trouble, I called my old instructor. To my surprise, he said that I had handled everything really well. He was surprised of my reports of the weather because he said he was sitting out on his patio watching all the jets going into Nashville. After further discussion, we came to the conclusion that it probably had something to do with the Tennessee River. We thought that the further I got from the river the better the weather would get. We agreed that after talking to Flight Service again and catching my breath, it would be safe to continue on to Murfreesboro. While looking at my chart I also realized that Centerville's visibility was probably a little lower than weather to the east due to the fact that Centerville was also near the Duck River.

I waited until the Centerville visibility rose to 5 miles, then I departed out to the east, heading home once again. Very shortly after takeoff the visibility greatly improved. Before crawling into bed after a mentally exhausting evening, I pulled out my weather book and looked over water effects on visibility!

The author of this "survivor story" is Aaron Hedman. Aaron is a former student of mine, but today he is a United Airlines B 737-300 First Officer. He wrote the story before the JFK, Jr. accident and it is used here with permission from Aaron.

I saw great similarities between Aaron's story and the NTSB report on the JFK, Jr. accident. Both pilots had little experience. Both were in that danger zone between 50 and 350 hours. Both received weather briefings that called for "good" weather. There was no reason to believe from either weather report that storms, low clouds, or poor visibility would be a problem. But despite the forecast, both encountered poor visibility over water. One landed as soon as possible. One lost control and hit the water. One survived this flight and went on to a promising career. One ended in a terrible tragedy.

I avoided an experience with IFR when I was a VFR pilot by deciding to stop before takeoff. It happened years ago. It was the summer of my sophomore year in college and I was visiting a friend in Alabama. I was supposed to fly home on Sunday night so I could get back to my summer job in Nashville. That Sunday the weather was terrible: low clouds, rain, storms. It was much more than my VFR private pilot experience could handle at that time, so I called my parents to help me out. If I waited and flew back on Monday morning I would miss work, and even though it was only a summer warehouse job to earn college tuition, I needed that job badly. Myself and the other summer employees had been warned by the foreman the previous week that if anyone missed work, they would be replaced immediately. I first talked to my mother on the phone. I asked her to call the foreman first thing in the morning and tell him that I was extremely sick and although I had really wanted to go to work anyway that "my mother" had overruled and forbidden me to go to work. My mother said, "well that's not really the truth, and I don't think I can call this man and lie." I said, "let me talk to Dad." I told my father that if the foreman ever found out that I was actually in another state and could not get back from a vacation trip, that I would be out of work before I got home. My father said that he would call the foreman and take care of the problem. I hung up. I flew home in VFR weather the next day. When I got home I asked my dad about his conversation with the foreman. He said that they had spoken and he expected me back at work the next day. The plan had worked. "Dad, what exactly did you say to him?" My father smiled and said, "I told him that you were 'under the weather.'"

I'm much older now and I do not advise that you lie to anyone, but if you can avoid a dangerous, potentially deadly, flight home by using my father's "under the weather" story—then be my guest.

Disorientation

What exactly is the problem with instrument meteorological conditions? Why is it so deadly? The real problem starts with the human body. We humans are perfectly adapted to our "home" environment—the solid Earth. But when we travel into our adopted environment—the sky, our Earth-bound senses can trick us. We can easily experience spatial disorientation. This phenomenon has also been called "pilot's vertigo" but vertigo is not an accurate term for this situation. Vertigo is a hallucination of movement. It is a sensation of rotation when there is truly no rotation. Spatial disorientation is different. When a pilot experiences this disorientation, is it because movement and rotation are present, but they are misinterpreted by the body.

Humans keep their balance and orientation with a combination of three sources of information. First, we use our eyes for balance. We do this when we sit, stand, and fly. I can level the airplane's wings by looking out at the horizon and simply matching the wings to that horizon. We understand what is up and what is down by seeing what is up and down. Our eyes are constantly feeding information to our brain about what is up and down without us ever even realizing it.

The second method that our bodies use for balance is called the *proprioceptic sense*. The tension on our muscles in our body assist in determining our position. This is sometimes called "by the seat of the pants." We are very accustomed to having the force of gravity act on our bodies from a certain direction. You certainly can feel that something is different when you are standing on your head. The brain receives signals from inside the body about how outside forces are acting on the body. The brain interprets these signals for balance and position.

Finally, we determine balance and position with the *vestibular apparatus*, which is a fancy name for the inner ear. As the name implies, the inner ear is located inside the head between the ear and the brain. Hearing is accomplished in the outer ear through the eardrum and into the middle ear. It is in the middle ear that the incus, malleus, and stapes bones are located. These tiny bones detect the vibrations that sound makes on the eardrum and transmits these vibrations to a fluid inside the inner ear. The middle ear is also the part of the ear that "pops" during altitude changes. The inner ear is actually two organs in one: one part for hearing the other for balance. The inner ear is hollow, but it is filled with fluid. The bones of the middle ear pass vibrations from the outside world through into the fluid. The vibrations create waves in the fluid that are detected by the temporal lobe of the brain and the brain interprets these waves as sound. The other part of the inner ear, also hollow but filled with fluid, are the semicircular canals. There are three such canals and they are positioned in three axes so that motion in three axes can be detected. Figure 3.3 illustrates the three canals and the ironic fact that the canals exactly align themselves with the pitch, roll, and yaw axis of an airplane. Who said humans were not meant to fly!

Together, the sources of information—eyesight, position of the body, and inner ear—combine to give the brain the big picture of what is going on. But what makes VFR flight into IMC so deadly is that when visibility is reduced by haze, clouds, rain, fog, smoke, or anything else, the horizon can no longer be seen and the eyes can no longer determine what is level with the horizon. So one-third of the source information to the brain is eliminated. The proprioceptic sense, or the position of the body, can easily get fooled. When on the ground and at rest the body can

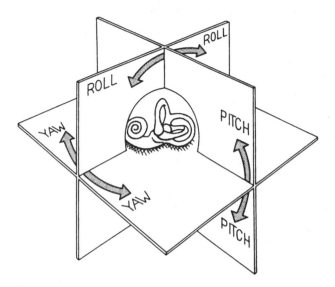

Fig. 3.3 The semicircular canals of the inner ear align with the pitch, yaw, and roll axis. FAA AM-400-90/1.

feel gravity pushing down and from that alone can determine what is up and what is down. But in flight there are other forces acting on the body like centrifugal force and acceleration. The proprioceptic sense can't tell the difference between gravity pushing down and centrifugal force pushing sideways so it becomes confused. The body sense in flight will quickly become unreliable. Of course these same confusion forces are also present during flight in visual conditions, but when the body incorrectly detects a turn, the eyesight can override that sensation by seeing that the airplane is in fact level. But when eyesight reference is lost, the body sense goes unchecked. In the clouds, eyesight reference is gone and body reference is unreliable.

The worst problem is with the last of the three—the inner ear. The inner ear is made up of the three semicircular canals. Each canal is hollow but filled with fluid. The canals detect movement and then pass this information on to the brain. The brain receives this information through sensory hairs that line the inside of the canals. Figure 3.4 illustrates just one canal and one set of sensory hairs. Illustration 3.4A depicts a sudden counterclockwise movement of the head and consequently the canal. Meanwhile, the fluid that was at rest in the canal momentarily remains at rest as the canal itself first moves counterclockwise. The fluid remains

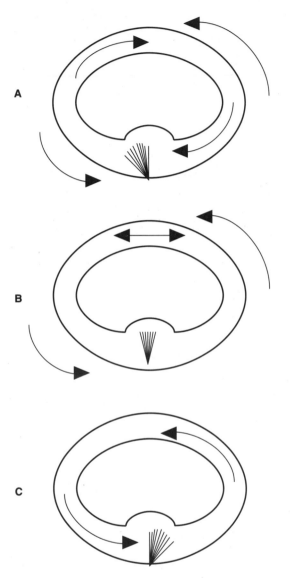

Fig. 3.4 The motion of the inner ear fluid in a single canal.

stationary as the canal is moving around the fluid. This creates the effect that the fluid is moving clockwise in relation to the inside of the canal. This will bend the sensory hairs in the direction of the fluid. When the hairs bend, a signal of motion is sent to the brain. This is how the brain "feels" when the airplane turns. The brain gets a sense of the direction of turn and the speed of the turn from the sensory hairs.

Illustration B of Figure 3.4 shows what happens just a few seconds later. As the canal starts to turn, friction between the canal's inner walls and the fluid starts to drag the fluid along for the ride. In this diagram the fluid inside the canal has caught up with the rotation of the canal. This means that less and less relative movement is taking place between the canal and the fluid. Eventually, even though the entire canal is still turning, the fluid within the canal is also turning at the same speed. The sensory hairs are not swayed and therefore stand erect. This signals the brain that no motion is taking place. But this is a trick on the brain! The airplane is still turning, but it feels to the brain that we have stopped turning. Disorientation means that you believe something that is not true. In this case the brain believes that we are not turning when in fact we are.

Illustration C of Figure 3.4 is the final killer. In this diagram the canal has been abruptly stopped, but the fluid continues on. It is like when you hit the brakes of your car hard. The car stops, but all the unattached stuff in the car (trash, lunch, unbelted people) flies forward. The stuff does not stop—it continues on. When an airplane turn is stopped, the fluid continues on and carries the sensory hairs along in the stream. This will send a strong signal of turning motion to the brain when in fact there is no motion. In these situations the inner ear sense of balance cannot be trusted. Remember the radar data from the John F. Kennedy, Jr. accident. Before the final plunge to the ocean, ... *"the airplane began a turn to the right and climbed...the airplane then started a descent and a left turn back to east. Thirty seconds into this maneuver, the airplane started another turn to the right and entered a rate of descent that exceeded 4,700 fpm."* The airplane turned right while climbing, then left while descending, then right again. At least two of the canals in each inner ear were sending incorrect information to the brain. There was no horizon to align with and override the turning sensations. The muscles and tendons were feeling movement but could not figure out from where. The inner ear was telling the brain that left was right and up was down. Reacting to the false information it became impossible to know how to fly the airplane and eventually the wrong control inputs were given for the true situation and control was lost. They went down not knowing which way "down" was.

Clouds and low visibility by themselves are not dangerous. But humans are not adapted to the loss of sensory perception that takes place inside a cloud. The challenge of instrument flight is to substitute the visible horizon with instruments that portray the horizon. Instrument pilots again use their eyesight to regain balance and position, but they use instruments in lieu of the real thing. While using the instruments a battle rages inside the pilot's body. The proprioceptic and inner ear senses will continually try to convince the pilot's brain that motions and accelerations are taking place that do not agree with the flight instruments. The pilot must ignore what the body is saying and trust what the instruments are saying—but this is very hard to do. After all, your body senses have been with you your entire life and have never let you down before. Your body senses have proven over time to be trustworthy during your life on the ground, but now you are being asked to toss out these loyal friends and place all your faith in a collection of dials and needles? When pilots fly VFR into IMC, this epic battle is fought. If pilots trust what has always worked, they probably will not survive the cloud encounters. But with practice it is possible to override the body sensations, trust the flight instruments, and cross the bridge into safe instrument flight.

The battle between what your body feels and what the instruments say is one that you will fight as long as you fly. You cannot "train-out" the body senses. No matter how much you fly, or what you fly, there will be occurrences where your body feelings will be in opposition to what is actually taking place. This threat will always be there. You cannot practice real hard and eliminate the problem; the only way to overcome it is to override it. Continued VFR flight into instrument conditions is deadly because the one thing you thought you could count on—your own body—turns against you. The balance senses that are normally actuated will either give no information (eyesight on the outside horizon is gone) or erroneous information (proprioceptic and vestibular systems are fooled) to the brain.

How can you safely experience spatial disorientation so that you can become a believer? The easiest way is to use a common chair that will freely rotate. You probably have one at the office or at home. Sit cross-legged in the chair with eyes closed and have a friend start you rotating. You will initially feel the turn's direction and speed (Figure 3.4A). If the chair can be continually and smoothly rotated without jerks, the inner ear fluid will eventually catch up with the turn and you will feel as if the turn is slowing down. When the fluid and the inner ear canals are moving together (Figure 3.4B) you will think that the turn gradually stopped when in fact you are still in motion. Now have the friend abruptly stop the chair (Figure 3.4C). Before the chair stopped, you thought you were already stopped, but as the chair is actually stopped, the inner ear fluid will rush on, creating a strong turning sensation. You will not feel dizzy with your eyes closed, but when you open your eyes a brain conflict will occur. With eyes closed the brain feels a turn that is in fact not taking place. When the eyes open the brain gets a signal from the eyes that there is no motion simultaneously with a signal from the inner ear that there is motion. You get dizzy as the brain tries to sort out the conflict.

This conflict is also a leading cause of motion sickness in both cars and airplanes. When you read in a car, there is no relative movement between your eyes and the pages of the book. The eyes therefore are sending a "no motion" signal to the brain. But meanwhile the body and the inner ear feel the bumps and sways of the car and send a "motion" signal to the brain. These conflicting signals to the brain makes some people sick. In an airplane if you get queasy while flying in VFR conditions, look outside so that your eyes, body, and inner ear will all send the same signal to the brain and you will probably feel better. Unfortunately, in IFR conditions your passengers will have no reference to override the queasy feeling. Instrument conditions and sick passengers can lead to a real bad flight.

Another way to experience spatial disorientation safely is to ride the Vertigon. The Vertigon is a simulator that is operated by the FAA's Civil Aeromedical Institute. FAA representatives take the contraption to air shows and wings weekends all over the country. Inside is an airplane cockpit with instrument panel and screen to show a view out the front of an airplane. Once inside the cockpit the door is closed so that the occupant cannot see anything outside. You guessed it; as soon as that door closes that whole contraption starts to spin around in circles. It even leans over sometimes while it's going round and round. You are guaranteed to get a taste of disorientation. Very seldom does anyone ever get sick in the Vertigon, but seat belts are required.

How long does it take to get spatial disorientation? The answer to that question would depend on several factors. It could be as quick as 20 seconds. The following are some sample NTSB accident reports from fatal accidents involving VFR flight into IFR conditions. The accident samples used here are representative of this entire class of accidents. The accident reports selected were done so at random. Please remember that the reason for examining an accident report is not to assess blame on any person, but rather to help us make better future decisions.

NTSB Number: FTW96FA071. South Padre Island, Texas

A non-instrument rated private pilot had accumulated a total of 215 hours of which 3.4 hours were simulated instrument time. The last simulated instrument time recorded was April 24, 1993 (the accident took place on December 16, 1995). Witnesses observed the airplane descend out of the "broken fog" at approximately 500–550 feet above the water. They reported that the airplane circled (360 degrees) as though trying to "avoid flying into fog." As the airplane turned toward a southerly direction, it began a gentle descent and slight left turn as it disappeared. The witnesses reported the engine was "running fine." Another witness then heard an explosion, and when he looked in the direction of the sound he saw what he thought was a boat breaking up and sinking. The witness proceeded to the area to search for survivors and recovered a wheel that was later identified as belonging to the accident airplane.

Probable Cause:

VFR flight by the pilot into instrument meteorological conditions (IMC), and his [the pilot] failure to maintain sufficient altitude (or clearance) above the surface of the water. Factors relating to the accident were: the adverse weather conditions (low ceiling and fog), and the pilot's lack of instrument experience.

NTSB Number: BFO93FA050. Ligonier, Pennsylvania

On January 15, 1993, the pilot requested and received a weather briefing from FSS personnel for a flight from Smoketown, PA to Latrobe, PA. The pilot was issued flight precautions and forecasts for IFR conditions. VFR flight was not recommended. A radar target was observed departing Smoketown at 0727 EST and was tracked towards Latrobe. The airplane was reported missing later that same morning and a search was initiated, but the airplane wreckage and occupants were not located until March 18, 1993. An examination of the accident site and airplane revealed that the airplane contacted trees at the top of a mountain and then impacted the terrain at an altitude of 2,800 feet MSL. No airframe or engine anomalies were noted.

The private pilot was noninstrument rated [194 total flight hours, of which 2 were simulated instrument flight] and no flight plan had been filed for the flight. Instrument meteorological conditions prevailed at the time of the accident.

Probable Cause:

Visual Flight Rules (VFR) flight by the pilot into Instrument meteorological conditions (IMC), and the pilot's failure to maintain sufficient altitude from the wooded terrain. The terrain and adverse weather conditions were related factors.

NTSB Number: SEA95FA031. Grantville, Utah

Localized adverse weather conditions, including low ceilings and snow were reported moving west to east as the airplane flew east to west. The non instrument rated private pilot [309 total flight hours of which 64 was simulated instrument flight] received a weather briefing for a VFR flight over 5 hours before he actually departed. He and his three passengers departed at night in mountainous terrain and in VFR conditions with intention of flying to an airport located 90 miles away for dinner. The pilot received ATC radar advisories and reported that the ceilings were getting lower along his route of flight. He was advised by ATC that areas of level 1 and level 2 precipitation existed in front of him. The airplane continued to descend after ATC services were terminated. Radar data for the airplane was lost shortly thereafter. The airplane impacted a mountain ridge about 6,200 feet MSL and was destroyed. The ridge was located along a direct line from the departure airport to the destination airport. No distress calls were recorded from the pilot, and no evidence of pre-impact mechanical deficiencies were found.

Probable Cause:

The VFR pilot's attempt to continue the flight into Instrument meteorological conditions, and his failure to maintain altitude/clearance with mountainous terrain.

NTSB Number: ATL83AA305. Elkin, North Carolina

During a weather briefing for the proposed route of flight, the pilot was advised that showers were expected in West Virginia and that it might not be possible to maintain VFR below clouds along the route. The briefer suggested an alternate route further east. The pilot, however, filed a VFR flight plan via the originally proposed western route. Witnesses heard engine sounds as the aircraft apparently approached a cloud. The airplane was not seen again until it spiraled out of the bottom of the cloud with one wing missing. [The pilot was a noninstrument private pilot with 242 total flight hours, of which 25 was simulated instrument time].

Probable Cause:

The VFR Pilot was seen flying into Instrument meteorological conditions, the airplane experienced structural failure as the pilot attempted recovery from disorientation.

These four examples took place in different parts of the country. They took place in different airplane types. Some were over flat terrain, others over mountainous terrain. But the pilots in these four examples all made one common mistake. They all flew their airplanes into instrument conditions. They were all noninstrument rated pilots with total flight time between 50 and 350 hours, and each was carrying passengers. After entering the clouds, each either lost control of their airplanes or flew into the terrain while attempting to stay out of the overlying clouds.

Solving this Problem

I often fly my university's executive airplane. We fly on athletic recruiting trips, and to conferences, ball games, and speaking engagements. When people first fly on the airplane, I always have a little chat with them. I say that there is no meeting, no ball game, no conference, that is so important that I will risk our safety to get to. I want that understood from the start because I would not want to be pressured into anything later. I told this to our university president (my boss) on his first trip. He said, "Paul, in this airplane, you can be the boss!"

This is an attitude that I think all pilots should believe and act on.

Remember the pilot and three passengers who flew into bad weather attempting to fly to dinner. I'm sorry, there is no dinner, no vacation, no family reunion, there are even no family medical emergencies that are so important that you should risk your life for. If you are not instrument rated and instrument proficient—don't fly into clouds or attempt to fly under clouds.

Three of the examples given took place in daylight conditions. This means that the pilots could see ahead and saw the clouds coming. But even though they saw the clouds they did not turn around. You should make every human effort to avoid flying VFR into instrument meteorological conditions, but if it does happen you should start a 180-degree turn back to clear air. It would be time well spent to practice a simulated instrument 180-degree turn maneuver with a flight instructor, but here are the basics of how to do it. At the first indication that a cloud has been entered or that visibility has been reduced below VFR minimums, glance at the heading and then start a shallow turn, either left or right. Do not exceed a bank angle of 15 degrees. You might be tempted to turn steeper in an attempt to get out of the cloud quicker, but a steeper bank will bring on additional load factor and altitude loss problems. While the shallow turn progresses, decide on a roll-out heading. This should be the opposite direction of the heading you glanced at before the turn started. Smoothly roll out on the opposite heading and be patient. You know that clear air was just behind you when you entered the cloud, so clear air will be in front of you after the 180-degree turn. At all times, keep a safe airspeed. While in the cloud and without a reference to the outside horizon, use the artificial horizon (attitude indicator). This instrument has a representation of the nose and wings of the airplane against a horizon. Just fly that little airplane in the instrument, keeping the nose on the horizon and the bank shallow. When you exit the cloud or when visibility starts to improve, continue flight away from the danger. Find the nearest airport and have dinner there.

Unusual Attitude Recovery

In the accident examples, the pilots did not turn away from danger when they had that option. But in each case they flew farther on. After entering IMC, two lost airplane control while in flight. One of these ripped a wing off when attempting to recover. To maintain airplane control you must maintain airspeed. Sometimes the only way to maintain or regain airspeed is to lower the nose and give away altitude. Of course, there are situations where you have no altitude to give. What then? It is a tough choice, but it would be better to contact the ground while flying the airplane under control than it would to lose control and strike the surface without control. Both would be extremely hazardous, but you would have a better chance while flying rather than not flying.

The recovery from an unusual attitude may vary some from airplane to airplane, but the standard procedure should be much like this:

> If the nose is high:
>> You are under a stall-spin threat, so lower the nose first.
>> If the wings are banked, level the wings after the nose is lowered.
>> Once the stall threat has passed, add power to reduce altitude loss.
> If the nose is low:
>> Bring the wings to level first, then raise the nose.
>> Raising the nose first would only tighten the turn.
>> Initially reduce power to prevent overspeed. Then add power to climb after the airplane is back under control.

After recovery from an unusual attitude, fly out of the IMC conditions. It might not be clear which way is out after the airplane has been through an unusual attitude. Just remember the heading you were last holding and go the opposite way.

Spin Recovery

Probably the most challenging flight maneuver is a spin recovery while in the clouds. In a spin, the airplane is moving faster than your ability to comprehend. This is why pilots do not respond or do not respond correctly to a spin. Their brain just can't keep up with what is happening. In the first part of this chapter, I made the case that you must trust your instruments, but in a spin all bets are off. The spin itself will render most instruments useless or inaccurate, leaving you not much to go on. While in a spin the altimeter will appear to "unwind." The vertical speed indicator (VSI) will show a down rate of descent that is at the maximum limit of the index. Most VSIs will indicate a maximum of 2000 feet per minute either up or down. The descent rate in a spin can exceed 6000 feet per minute, so the VSI just can't keep up. The airspeed indicator in a spin will show the slowest possible speed. Most airspeed indicators do not go all the way down to zero, but instead have a low-speed peg. The needle will be pushing against that peg. The attitude gyro and heading indicator, being gyro instruments, will probably have tumbled. In this case these two instruments will sort of swim around in no particular direction and will provide no helpful information. The ball of the inclinometer, which is usually quite reliable, will always displace out to the left when the inclinometer is mounted on the left side of the instrument panel. Regardless of whether the spin is to the left or to the right, the ball will sway to the left because centrifugal force from a turn in either direction will toss the ball out. Just for proof of this, I have a separate inclinometer that can be mounted on the right side of the instrument panel. In a spin of either direction, that right-side-mounted ball swings out to the right, while at the same time the left-side-mounted ball swings out to the left. I use this when I teach spins to prove that you cannot use the old saying "step on the ball" in every situation.

The only instrument left then is the turn coordinator or, if you have one, the turn and bank indicator. The altimeter tells you are going down. The airspeed indicator, when it reads the lowest speed, tells you are stalled. The turn coordinator tells you that you are in a spin and in which direction. So it is the turn coordinator that tells you which rudder to use to stop the spin—the rudder opposite to the direction of the turn. Airplane spin recoveries vary, but here are the basics:

Reduce power to idle.
Hold *opposite* rudder to stop the spin.
Give brisk *forward* elevator to break the stall.
After airspeed is regained, raise the nose.

After the spin the gyro instrument will most likely still be tumbled and unusable. This will make it harder to maintain level flight, so as a backup, use the pitot-static instruments. Move the controls so that the pitot-static instruments stop their movements and reverse their trends. In other words, raise the nose smoothly until the VSI stops showing a descent and starts showing a climb. This will stop an ever-increasing speed shown on the airspeed indicator and reverse an altitude loss shown in the altimeter.

You may have heard that if you are in a spin, you can recover by simply releasing your controls. The strategy here is that if you don't know which way is up or what is going on, it would be better to just let go than to accidentally make the wrong control inputs. Remember the pilot who was disoriented and then created pilot-induced wing failure. That pilot obviously did not use the proper control input for recovery, and it may very well have been best in that case to just let go. The problem is that not all airplanes will recover by just letting go. The airplane that I routinely conduct spin training in will indeed recover on its own, but only if I let go before the third turn. If I let go after the third turn, it does not recover on its own, but instead continues to spin! That would mean in order to use the "let go" recovery method, I would have to be thinking fast. Three turns take just over three seconds in that airplane, so I have less than three seconds to remember to let go. Other airplanes will recover after letting go regardless of the number of turns; talk to a flight instructor who has spin experience in the airplane you fly.

The reason a spin recovery is so unlikely from IMC is that it does require fast, smart action from the pilot. The problem is that if the pilots were displaying fast, smart action, they would not have flown into the clouds in the first place. It is much to ask for a pilot to go from not-so-smart to incredibly smart in a split second while under maximum stress. This is why weather accidents have the highest death rate.

I have a friend who is an investigator at the National Transportation Safety Board. He says, "If you have time to spare—go by air." He means don't ever place yourself in a situation where you must get somewhere and then try to use an airplane to cut down the travel time. You must build in spare time because if the weather gets bad, you must wait. To be completely safe, pilots should be required to carry their pilot certificate, their medical certificate, and a standing account at a rental car company.

Flight Instruments

It must be again stressed that the best prevention of a weather-related accident is to avoid the weather all together. But trusting the flight instruments is a requirement for even student pilots. Today, three hours of flight by reference solely to flight instruments is required for the private pilot certificate. Some of the accident cases discussed earlier involved pilots whose total instrument experience were those hours required for their private certificate. Those hours are not adequate for actual instrument flight, and pilots should not try to convince themselves otherwise. Those hours are training intended for the emergency 180-degree turn from the clouds back to safety, not for cloud penetration on purpose.

But no matter your level of instrument experience, pilots must trust the instruments and be ready to troubleshoot them. We have said that in the clouds two of the three sources of balance information (proprioceptic and vestibular) become completely unreliable. The only remaining source of information left to save your life is eyesight. But the eyes cannot be used to level the wings against the real horizon when in reduced visibility or clouds, so the eyes must shift to the flight instruments. The flight instruments, working together, will present to the pilot's eyes a picture of the attitude of the airplane and what is going on. The pilot must interpret the picture and then use this information to pilot the airplane safely. You can see that a misinterpretation could lead to disaster.

As an instrument student, you will spend much time working on your "scan." The instrument scan is the way in which you move your eyes across the various instruments in an attempt to get pertinent information, get the big picture, and fly the airplane accordingly. I believe that the scan is a very personal thing, and that you will develop your own scan with practice. The scan can not be standardized. A predetermined scan pattern will not be adequate in every situation. In a turn, the turn coordinator will require more attention than in straight-and-level flight. In a climb, the vertical speed indicator will demand more consideration than when flying level. The scan changes with the situation. I cannot suggest a path for your eyes to take as they look across the panel—it is one of those things that you know it when you see it—but it takes practice and then more practice.

Instrument Failure

As our eyes flash across the instruments gaining insight and making flight corrections based on this insight, we take one thing for granted: The instruments are telling us the truth. But it is possible for a flight instrument or even a "family" of flight instruments to fail. If an instrument fails, it will likely start to give false information to the pilot. Our defense against disorientation is the information from the flight instruments, so if this information becomes a lie, disorientation again becomes a death threat. If one or more instruments stops telling us the truth about what the airplane is doing, the pilot must quickly:

1. Identify which instrument is lying.
2. Ignore the instrument that has failed.
3. Start using a combination of functioning instruments to avoid disorientation.

The most insidious instrument failure, and therefore the most deadly, is the vacuum system failure. Traditionally, the vacuum system operates the attitude gyro and the heading indicator. Figure 3.5 is an illustration of the principle of the vacuum system. The vacuum pump is usually engine driven and therefore mounted to the engine itself with a gear through a shaft. When the engine turns the pump turns and air is drawn through the system. Some systems arrange things so that air is blown through the system; check your manual or ask your A&P technician. In Figure 3.5 air is drawn through the system just like sucking through a straw. Air comes in through a filter and then is routed through both gyro instruments. This diagram illustrates the operation of just one of the gyros. Air is directed across the gyro wheel. The wheel itself has buckets, just like an old-fashioned water wheel, that cause the gyro to turn. The faster the wheel spins, the more rigid in space it becomes and the more trustworthy the attitude gyro and heading indicator become. The speed of the gyros is monitored by a suction gauge (not shown) that tells the pilot how much differential pressure the vacuum pump is producing. The suction gauge usually has a green arc to display adequate air flow to spin the gyros. A serious problem can occur when that vacuum pump fails. The connection between the engine and the pump is a shaft that ironically is designed to fail! If there ever is a binding of the pump, the shaft will easily twist off saving the pump from further damage, but leaving the pilot with no gyro suction. If this happens the air flow will stop, but the gyro itself will not stop instantly. The gyro will spin down gradually, and as it does the rigidity of the gyro will be lost gradually. The attitude gyro and heading indicator will fail slowly, and this can trick and mislead the pilot.

To pass the instrument rating practical test, a pilot must complete at least one instrument approach with no gyros. This is to simulate the vacuum system failure and prepare for it. In training, instructors will routinely simulate the failed vacuum system by placing a cover over the attitude gyro and heading indicator and this is good practice,

VACUUM GYRO PRINCIPLE

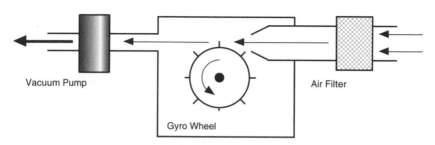

Fig. 3.5 Vacuum gyro operating principle.

but it is not "real" practice. You see, when the instructor covers the instruments you know to immediately look at other instruments; the detection of the failure is done for you. Also, this simulates an instantaneous failure rather than a slow failure. Flying without vacuum gyro instruments is one thing; detecting their failure as they die gradually is another.

Under normal circumstances, the flight instruments should complement each other. In a left turn, the heading indicator should show heading changes to the left; the turn coordinator and attitude gyro should show wings banked to the left. If ever you suspect a disagreement among the instruments, then you must immediately become a detective. The airplane cannot both climb and descend at the same time or turn both left and right and the same time, but someday your instrument might try to tell you differently. When a conflict occurs, you must quickly identify the instrument that is lying, ignore that instrument(s), and switch to information from other instruments to avoid disorientation. Ordinarily the turn coordinator is electric and does not need any vacuum air to operate. This is the "don't place all your eggs in one basket" theory. It would be very rare to have both an electrical and vacuum failure simultaneously, so if conflicting information is presented it is a failure of one or the other. If the attitude gyro is showing a turn but the heading indicator is not in motion, then you have a vacuum failure. Check the suction gauge to confirm this and shift your attention to the turn coordinator and magnetic compass. Most turn coordinators have a red flag that becomes visible as its gyro spins down. Check this flag to confirm the operation or failure of the turn coordinator.

Vacuum system failure is scary and will catch pilots who fixate. Fixation is a scan problem whereby the pilots fix their gaze on only one or two instruments. If the pilot does not cross reference all the instruments, but instead stares at just one, and that one fails, then the pilot will be receiving only false information and will not be able to detect a problem until it is too late. Fixation on the attitude gyro is common because it gives the most real picture of the horizon and bank angles, but the attitude gyro is the first to go in a vacuum failure. Soon after vacuum failure, the artificial horizon of the attitude gyro will slowly start to show a turn. Flying with this instrument alone, the pilot—thinking the slow turn is actually taking place—will roll the airplane to stop that turn. But in doing so the airplane starts a turn in the opposite direction and reality is lost; airplane control will be soon to follow.

When you fly without vacuum instruments, you should make long, shallow, slow turns because you will only be left with the magnetic compass for direction information. The mag compass will lead and lag, and it is very frustrating to use even under smooth air conditions. The regulations call for you to advise ATC whenever a system failure occurs, such as a vacuum system, but there is not too much help they can give. A fatal accident took place once following a vacuum system failure. The pilot detected the problem and reported to ATC. "I have a vacuum system failure," the pilot reported, but the air traffic controller did not know what a vacuum system was or what the nature of the problem actually was. The controller sought the help of another controller. Controllers can speak to each other on a land line without the pilot hearing their conversation, but all communications are recorded. The first controller said to second controller, "Hey, I have a pilot who just reported that his vacuum 'cleaners' just went out." The second controller said, "OK he has no instruments and will need to get down fast!" The controllers began assigning heading changes to direct the pilot quickly to the nearest instrument approach. But the airplane was not in danger of falling out of the sky as long as the pilot kept his orientation. Getting on the ground as soon as possible was not as important as maintaining control. The last thing the pilot needed was a series of abrupt heading and altitude changes. The controllers thought the best remedy was to expedite, when in fact the pilot needed slow, long, shallow turns to maintain control. Unfortunately the pilot took the controller's advice, made a series of sharp turns, became disoriented, and crashed. I have had three IFR vacuum system failures in 25 years, and they are a real challenge. If you ever have a vacuum system failure, remember a good scan leads to detection. Detection leads to solution. Advise ATC, but stay in command of your airplane. You want a long, easy, narrow intercept vector to the final of an instrument approach. Reject anything else.

NTSB Number: FTW96FA046. Cypress, Texas

The pilot received three separate weather briefings of forecasted IFR conditions in the six hours prior to his departure. While flying in instrument meteorological conditions en route to Conroe, Texas, the pilot reported to Houston Approach Control that, "I lost my gyros." No-gyro vectors were given to the pilot by Houston Approach Control in an attempt to align the airplane for an ILS approach to runway 8 at Houston Intercontinental Airport. The pilot lost control of the airplane and impacted the ground. The weather conditions at David Wayne Hooks Municipal airport which is 10 nautical miles north east of the accident site, at 1850 CST were estimated ceiling 500 feet AGL overcast, visibility 1 mile with fog, temperature of 57 degrees, dew point of 56 degrees with wind at 070 degrees at 6 knots, and an altimeter setting of 30.17 inches. Examination of the vacuum pump input shaft by NTSB revealed that the shaft failed as a result of sudden stoppage of the vacuum pump.

Probable Cause:

The pilot's failure to maintain aircraft control while flying in instrument meteorological conditions. Factors were the weather, and failure of the airplane's vacuum pump that rendered the attitude indicator and directional gyro inoperative.

The gyro instruments form one major "family" and the pitot-static instruments form the other. Figure 3.6 is a typical pitot static system. The pitot tube and static ports are mounted outside the airplane and are inspected as part of a routine preflight check. The instruments in this family are the altimeter, vertical speed indicator, and airspeed indicator. The airspeed indicator is the only instrument that uses both "ram" air from the pitot tube and "static" air from the static ports; the other two use static air only. Static pressure is the weight of the air pushing down on us by gravity. Air is very compressible, so gravity is able to squeeze most of the air down toward the surface. This is why the air becomes lighter (less dense) as you climb. On a standard day, the weight of the air is 14.7 pounds per square inch at sea level. That means that over our entire body, we have a couple of tons of air pressing down on us. We have a hard time actually feeling this air because the air pressure is not only *on* our body, but it is also *in* our body. The pressure is equalized and we live our lives without noticing the pressure much. Notice it or not, it can be measured and it is one of the two forces that operate the pitot-static system. The other force is ram air. This is the force of air you feel if you stick your hand out the window of a moving car. The faster the car is traveling, the greater pressure is felt.

Instruments of the Pitot-Static family can also give incorrect or misleading information to the pilot. To understand how to troubleshoot the system in an attempt to maintain orientation, you must understand what makes the system work. Figure 3.7 is just the part of the system that operates the airspeed indicator. Figure 3.7 is absolutely not a technical drawing, but instead "concept" drawings so we can see what is happening. In diagram A of Figure 3.7, we see normal operation. Ram air comes in through the pitot tube and essentially inflates a chamber inside the airspeed indicator. The ram air chamber is separated from the static air chamber by a movable diaphragm. The static air attempts to hold back the diaphragm, but the ram air is much stronger and pushes the diaphragm back as shown. The amount of bulge from the diaphragm is relayed to the needle on the face of the airspeed indicator by a series of cogs and gears. The more bulge, the faster airspeed reading.

Great care must be taken to ensure that all of these "air openings" are clear and open before flight. Insects have been known to use these holes as nests. If dirt and debris are detected in the pitot tube, do not try to clear the blockage by inserting anything into the tube. Doing so would just push dirt further into the tube. The air inside the pitot tube runs through a plastic hose. The hose has a fitting that connects the pitot tube to the airplane. This fitting must be opened and high-pressure air blown from inside out to clear the debris. This is why you should always use a pitot tube cover and also why you should always remove it before flight.

But in flight it is possible for one or more of these openings to get clogged. Ice can form over the entry hole or undetected debris inside

PITOT - STATIC SYSTEM

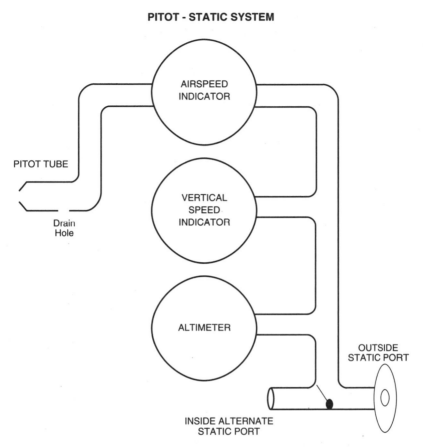

Fig. 3.6 Pitot-static system.

can become dislodged by the in-flight airflow and clog the tubes downstream. Illustration B of Figure 3.7 shows a situation where the pitot tubes ram air entry has become blocked. In this situation the airspeed indicator will read zero—or whatever its lowest speed might be. Look carefully at the diagram. Downstream of the pitot tube's entry port is another hole—a drain hole. Normally the drain hole allows water that comes into the pitot tube to drain out. Unfortunately, when air comes in, sometimes water comes in as well (rain). The air enters the tube and then is directed through a sharp turn up into the wing. The water, being heavier that air, can't make the turn and flows out the drain. This keeps water out of the system and away from the instruments. It should be noted that large amounts of rain can still get past the turn and get into

AIRSPEED INDICATOR PRINCIPLE

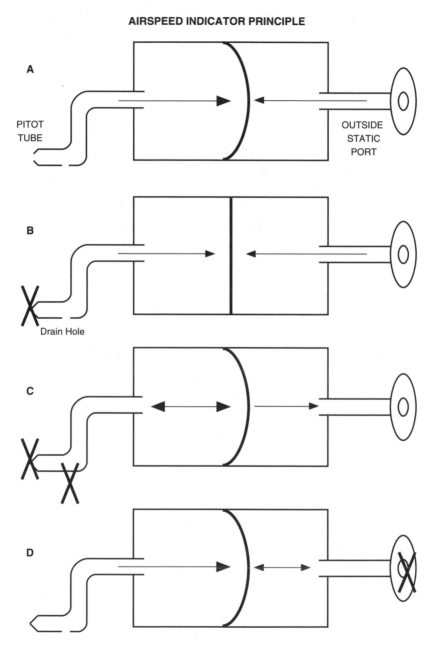

Fig. 3.7 Airspeed indicator operating principle.

the system. When flying in rain, or even flying after the airplane has been sitting out in heavy rain, you might see a sudden jump on all the pitot-static instruments. This simultaneous jump can be due to water working its way through the system and momentarily blocking the air pressure readings. It will not last but a second, and then it should return to normal. Many pitot-static systems have manual drains that must be opened periodically to ensure that water is removed from the system. Now the drain hole plays a part in the malfunction that occurs when the ram air becomes blocked. When the ram air disappears because of the blockage, the drain hole acts like another static port. You can see from illustration B that air of equal pressure can now come from both sides. Air from the outside static port arrives at one side of the diaphragm and air from the drain hole arrives at the opposing side. Both have the same outside pressure so the diaphragm will not bulge in either direction. No bulge means no airspeed reading.

If the blockage is due to ice accumulation, you should turn on the pitot heat if you have it. This should melt out the ice and ram air will soon inflate the airspeed indicator again. Of course if you have pitot ice, you may also have other problems, such as structural ice on the entire airplane. Get to warmer air immediately and leave the pitot heat on. Diagram 3-7B is also what the system would look like if you went flying with the pitot tube cover on. I don't think that trying to "burn off" the cover using the pitot heat will work and it is not a very good idea! The good news is the situation in illustration B is easily detected. The airspeed will abruptly drop to zero; you can't miss it.

Illustration C of Figure 3.7 shows a more sneaky situation where both the ram air and drain hole are blocked. When this happens the "airspeed indicator will act like an altimeter." This means that if you climb, the airspeed indicator will show an increase in speed and if you descend it will show a reduction in speed. Higher altitude/higher numbers, and visa versa. Now this could slowly catch a pilot off guard because there will be no abrupt change in airspeed indication.

When both ram and drain holes are blocked, it's like blowing up a balloon and then sealing it off with a knot. The previously inflated air becomes trapped inside. This will then hold a constant amount of pressure against the diaphragm. If the airplane starts to climb, the pressure outside will become less and the air on the static side of the diaphragm will escape out the static port. This means that there is now less static pressure to push back the ram air. With less pressure to hold it back, the pent-up ram pressure bulges the diaphragm farther and this moves the airspeed indicator's needle to a faster value. How can a pilot detect this? Well, the airplanes I fly cannot both climb and accelerate at the same time. If I see a climb on the VSI and a simultaneous increase of airspeed, I know something is not right. On the other hand, if the airplane descends, more static pressure will enter the static side. This provides additional force to push back the ram air side. This means less bulge of the diaphragm and less speed shown on the indicator. Apply pitot heat anytime you suspect an instrument conflict between speed and climb.

This "airspeed acts like altimeter" condition was the cause of a 1974 accident of a Boeing 727. The charter airplane took off from New York's Kennedy airport bound for Buffalo, New York, where they were to pick up the Buffalo Bills football team. The airplane never made it to Buffalo. There were no passengers on the airplane, just the crew of three experienced pilots. Climbing through 23,000 feet and in IMC, the first officer noted that the airplane's rate of climb was 6500 feet per minute with an indicated 405 knots of airspeed. This was excessive even for an empty jetliner. The cockpit voice recorder picked up the conversation from there:

Captain: "Would you believe that #@&X$."

First officer: "I believe it, I just can't do anything about it."

Captain: "Pull her back and let her climb."

Soon after the first officer pulled back on the yoke, the stick-shaker (stall warning) went off. The airplane stopped climbing at 24,800 feet with an indicated 420 knots. The airplane was on the edge of a stall, but the crew trusted what the instruments said and therefore believed that they were too fast rather than too slow.

First officer: "There's that Mach buffet, I guess we'll have to pull it up."

Captain: "Pull it up."

The mach buffet is a vibration, formed by shock waves, that takes place when the airplane exceeds it critical mach number. The crew never entertained the idea that the speed was slow and that the airspeed indicator was in error. They never scanned other instruments. They never thought that accelerating in a climb was suspicious. Then the landing gear warning horn sounded, indicating that the throttles had been retarded with the gear up. Thirteen seconds after arriving at

24,800 feet, the aircraft began to fall at a rate of 15,000 feet per minute while turning rapidly to the right. The airspeed was now decreasing at a rate of four knots per second.

Unidentified crew member: "Mayday! Mayday"

New York controller: "Go ahead…"

Unidentified crew member: "Roger, we're out of control…descending through 20,000 feet."

New York controller: "Altitudes below you are clear of traffic."

Unidentified crew member: "We're descending through 12—we're in a stall."

That was the final radio transmission. But inside the airplane the conversation continued.

Captain: "Flaps two!"

First officer: "Pull now—Pull that's it."

The vertical acceleration of the airplane was greater than 5 g's. At 3500 feet, the left horizontal stabilizer separated from the airplane, and at 1090 feet MSL, the airplane hit the ground. They had fallen over 23,000 feet in 83 seconds.

The NTSB concluded later that moderate airframe icing was present during the flight above 16,000 feet. The crew had received an adequate weather briefing that included the icing conditions, but the pitot heater switch was found in the off position. The NTSB listed the probable cause as ice blockage of both the pitot tube ram inlets and drain ports, creating an erroneously high airspeed indication.

Trusting untrustworthy instruments is not just something that beginners get caught doing. It can trick even a veteran who is not suspicious. When it comes to flight instruments, we must "trust, but verify."

Illustration D offers one more problem. This time the ram and drain holes are open and the static port is closed. This condition will create incorrect airspeed indications. The only time correct readings are shown is when the ram air and static air come from the same altitude (same density). If the static port is blocked, the air on the static side of the diaphragm becomes trapped. The static side now has a constant pressure which is equal to the pressure at the port when it first became blocked. Any climb or descent will take the airplane to where the pressure in the static side is different from the real outside pressure. Any change in altitude will make airspeed indications inaccurate. Now inaccurate airspeed readings are not necessarily a dangerous thing as long as you realize the problem exists. The bigger problem regarding a static port blockage lies within the other instruments of this family.

The static side of the pitot-static system is used by all three instruments in this family. Figure 3.8 illustrates the other two: the vertical speed indicator and the altimeter. Again, these are "concept" drawings, not technical drawings, but they are used here for simplicity and to get the point across. Both VSI and altimeter also have flexible diaphragms. They are actually aneroid barometers that are sensitive to pressure changes. The VSIs diaphragm has a "calibrated leak," which is a hole so that no pressure can be contained for very long. If pressure does change, in time the higher pressure will push through the hole and eventually equalize. Illustration A of Figure 3.8 depicts the airplane in a descent. As we go down deeper in the atmosphere, more molecules of air will enter the system. These molecules push on the diaphragm of the VSI and bulge it in the direction of the stronger force. This bulge is shown on the face of the VSI as a descent indication. If we stop descending, the air pressure will soon equalize through the hole. With no unequal pressure there is no bulge and no climb or descent shown on the indicator. Meanwhile the altimeter shows a descent when additional air pressure comes in through the port and squeezes back a sealed container of air. The container is like a bellows that can expand or contract. It contracts during descent as more air pressure from lower altitudes acts on the bellows.

Illustration B of Figure 3.8 depicts the airplane in a climb. The air is thinner the higher we climb, so molecules are escaping the system out the static port. This bulges the VSI diaphragm and a climb is indicated by the needle on the face of the dial, and the airspeed indicator's bellows is expanding because less static air is present to hold it back. All this is normal operation, but what happens when the outside static port becomes blocked?

We have already discussed that the airspeed indicator will begin to read incorrectly once any altitude change occurs. Illustration C of Figure 3.8 shows what happens with the other two. With the static port blocked, there will be no pressure changes for the VSI and altimeter to detect. With no differences to measure, both instruments will "freeze." The VSI will show no climb or descent regardless of an actual climb or descent. The altimeter will read the altitude that it was at when the blockage occurred and will not change thereafter. It would look like you

VSI and ALTIMETER PRINCIPLE

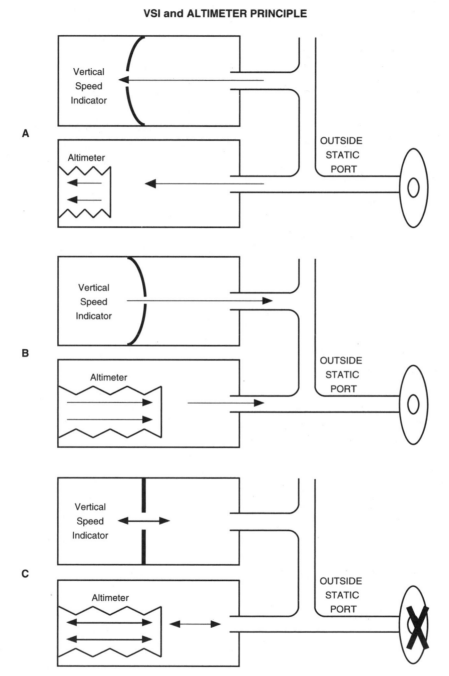

Fig. 3.8 VSI and altimeter operating principle.

are holding altitude better than you ever did before! I think I'm a pretty good pilot, but I cannot hold altitude so tightly that the needles do not move. I know that if they completely stop moving it's not because of exceptional skill; it's instead blocked static ports.

Some airplanes have heated static ports to melt out ice just as a pitot heater would. Some airplanes have more than one static port so that pressure can equalize and so that at least one might remain open. Still other airplanes have an alternate static source. Figure 3.6 depicts such an alternate. These systems have a valve that under normal operations directs the outside static air into the system. But should that opening become clogged, the valve can be reversed to allow static air into the system another way. The alternate air opening is usually inside the cockpit if the airplane is not pressurized. If the airplane is pressurized, the opening could not be inside the cabin because then the altimeter would read whatever the cabin was pressurized to. In that case the alternate static air opening is inside a wheel well or someplace else that is less likely to obtain ice.

Using inside alternate static air in a nonpressurized airplane obviously means that air going into the system is coming from inside. Taking air from inside will also have an effect on the pitot-static instruments. When air flows around the airplane's fuselage, it separates and accelerates just like over an airfoil. This means that the airflow around the cabin creates a lower air pressure around the cabin. Unpressurized airplanes are not airtight, so the slightly lower pressure just outside draws air out of the cabin, making the inside pressure slightly less than outside. Then when we use this slightly lower pressure air in the alternate static port, we are sending slightly lower air pressure into the pitot-static system. When the system senses slightly lower air pressure, the altimeter will read slightly higher, the airspeed will read slightly faster, and the VSI will show a momentary climb. But none of these errors will be dramatic. In fact, with slow-moving airplanes the difference in air pressure from inside and out may produce an error that is too small to detect. The alternate static valve should be tested before each IFR flight. You will know that air pressure is changing by the jump in the pitot-static instruments when you move the valve.

In order to maintain orientation, all of these factors must be considered because we must trust the instruments. But we should only trust those instruments that are trustworthy. It is up to the pilot to compare the instrument readings and troubleshoot the problems that might occur. Many fully IFR rated and cloud-proficient pilots have trusted instruments that were not trustworthy, and became disoriented. Disorientation, even among instrument pilots, is deadly.

Weather Wise

In each example of VFR pilots who fly into IMC, there came a point when the pilot seemed trapped. The weather was worse than forecast, but believing what they wanted to hear, they pressed on hoping that the promise of better weather was just ahead. Remember the weather reports the night of the JFK, Jr. accident. It was 6 miles visibility, sky clear, and wind 10 knots. It sounded like a perfectly good VFR evening. But the actual visibility, especially over the water, was much worse. The flight progressed with the thought of good weather and no problems but came across something unexpected.

We all want the weather to be great and the flight to be fun and uneventful. We want this so badly that sometimes we hear what we want to hear from a weather report. Then later, during the flight, if the weather is worse than expected we talk ourselves into believing that everything is really okay, and that this is simply a bad spot. We condition ourselves to the positive and we hold that belief even when their own eyes have evidence to the contrary. We don't want to believe that we have made a judgment error by beginning the flight in the first place. We see what we want to see.

Only a good thunderstorm seems to wake us up. VFR pilots who are foolishly unafraid of IFR conditions, thankfully are afraid of thunderstorms. There are very few accident examples where pilots have flown into thunderstorms. Pilots know the extreme dangers of a storm and there is never any question about pressing on. Storms universally mean stop. But what we have learned is that the danger is not storms but instead the seemingly benign clouds or low visibility that pilots encounter.

Pilots should, of course, learn to get and use weather reports, but we must do more. Pilots must become their own weather observers and forecasters. Pilots need to do a better job of looking out for themselves. Every weather-related accident could have been prevented if the pilot had done a better job of seeing what was coming. Here are some tips to help you see into the weather future of your next flight.

Haze

Haze can reduce visibility and hide the horizon. It is especially dangerous because haze is not often reported and rarely will appear in a forecast. This makes it possible for a pilot to be fooled by haze. Haze is a concentration of very fine, dry particles. These particles are usually invisible to the naked eye up close. This means that weather observers on the ground, whose vision is limited because they are on the ground, can be in a haze layer and not see it. If they can't see it, they will not report it. So haze can be present but not show up in any weather report or forecast.

Here is where the pilot must outsmart the weather reports by knowing the other conditions that must be present for haze to exist. Remember, haze is a "concentration" of particles. The atmosphere must be stable and the winds light for a haze layer to form. If the atmosphere were unstable, the microscopic particles would be mixed, churned, and dispersed in all directions. The haze particles must congregate in order to reduce visibility so when the particles disperse the problem is solved. Watch out for days and nights when the atmosphere is stable and winds light. How can you determine the stability of the air? One way is to ask the FSS weather briefer for the "lifted index." On a standard day, the air cools as you get higher in the atmosphere. Near sea level the weight of the air, pressed down by gravity, compresses. There are more air molecules per cubic foot at sea level just because gravity squeezes the air to the surface. This makes the space inside the cubic foot crowded with molecules. The molecules "rub elbows" so to speak, and this causes friction. The friction causes heat. It is warm at sea level due to this compressional heating. At high altitude there are fewer air molecules in the same amount of space. This means that the molecules have more room and create less friction. Less friction means less heat, and so usually it gets colder as you climb. The lifted index measures the atmospheric stability by comparing how quickly or slowly the air cools off with altitude. The standard cool-off rate for dry air is 3 degrees Celsius per 1000 feet. At that rate if you were to take standard air at sea level and lift it to 18,000 feet, it should cool off 54 degrees (3×18). When the lifted index is a positive number, this means that if the sea level air were lifted to 18,000 feet, it would be cooler than standard; it cooled off more than 54 degrees. This air, being cooler and therefore heavier, would sink.

Sinking air is stable air. A negative lifted index number would mean that the lifted air would be warmer than standard. That air, being warm and therefore light, would rise like a hot-air balloon. Rising air is unstable and is one of the ingredients of a thunderstorm.

Just ask the weather briefer for the lifted index on your next FSS call. Also, many computer and Internet weather sites now offer the "lifted index." If the lifted index is a positive number, then the threat of thunderstorms will be very low, but the possibility of haze will be very high. As the air sinks it joins together all those particles that alone are no problem, but that in concentration can lead to pilot disorientation. A positive lifted index number, together with light winds, on a night flight can come together to make a VFR forecast an IFR reality.

Fog

Unexpected fog can quickly reduce visibility and trap both IFR and VFR pilots alike. A fog is the same as a cloud. The only difference is that a cloud that is considered a fog has its base on the ground. A cloud has its base somewhere above the surface. All clouds and fogs form the same way: The air must become saturated with moisture. Here is a quick review of how that takes place.

We take water for granted. We use it every day in many ways, but we forget that water does have some unusual properties. The H_2O molecule is the only molecule that can be found in all three states (gas, liquid, solid) at normal temperatures found on Earth. When it is liquid we call it water. When it is solid we call it ice. When it is a gas we call it water vapor, but water vapor is often misunderstood. Water vapor is always invisible. Steam is not water vapor because it is visible. Steam, clouds, fog, are all liquid, but in these cases the microscopic liquid H_2O molecules are too light to fall and remain suspended in the air. But a cloud is liquid. If you can see it, then it is not water vapor. In the room where you are reading this, there is probably gallons of H_2O all around you in the room, but you can't see it because it is currently in the gaseous state. You can't see it but it is there. How can we get the H_2O to change state from a gas to a liquid so we can see it again? For the molecule to change its state we must bring the air in the room to a saturation point.

In any space there are a certain number of dry air molecules (oxygen, nitrogen, etc). If we start to add to the mix a bunch of H_2O molecules, there will come a point where the dry molecules can't make any more room for the H_2O molecules. When the air gets completely crowded, it is said to be completely "saturated." If one more molecule of H_2O is added, it's like a glass overflowing; there is simply no more space and the H_2O molecule is kicked out. This changes its state to a liquid and a cloud forms. This is also called condensation. The percentage of saturation is the relative humidity. 100% relative humidity would equal saturation. So a cloud can form and visibility can be reduced when too much moisture is added to the air.

This is what happens over bodies of water. Water will gradually evaporate from the surface of the water. This adds water vapor to the overlying air, increasing its humidity and pushing it closer to saturation. This can produce fog right over the water's surface while just a short distance away there is no fog over dry land. Fog can also form when rain falls to the surface.

There is another way fog can form. Rather that the water vapor content going up, fog can form if the temperature goes down. When the air is cold the air molecules have less energy, they become lazy, and they pile up on each other. This makes cold air dense. When the space is already dense with cold air molecules, there is less additional room for H_2O molecules. When there is no more room, saturation is reached and condensation occurs. The water vapor content can remain the same, but fog will form whenever the air gets cool enough to become saturated. The temperature where saturation occurs is called the dewpoint. This is another misunderstood term because it really has nothing to do with dew. Morning dew is liquid drops that form on your car, grass, plants. The dewpoint is the temperature where clouds (liquid suspended in the air) form.

Pilots can predict fog by looking at the difference between the actual temperature and the cooled temperature that the air would have to be to become saturated. This is the temperature/dewpoint spread. Fog can be expected anytime the spread is four degrees and less. A pilot planning an early morning flight should know that the temperature usually rises as the sun gets higher in the sky. A 4-degree spread at 7:00 a.m. means that fog is less likely to form because the temperature is getting hotter and therefore the spread is getting wider. The fog is said to "burn off." But a 4-degree spread at 7:00 p.m. is trouble. As the sun disappears there will be no more direct heating from the sun and the temperature will drop. The spread is already only 4 and will probably narrow. The weather-wise pilot knows that fog formation is just an hour or so away. Ask for area dewpoints when planning any flight and anticipate temperature changes.

On the coastline, moving fogs will form. When wind carries warm, moist air over a cooler surface, the combination of moist (high dewpoint) air and dropping temperatures can instantly create low clouds and fog. If the water surface is warm and the wind carries the air over cooler land, the fog will stream inland. Ever been to San Francisco? The bay area is a working model of dewpoint spreads and H_2O state changes. As a pilot, look for onshore wind forecasts and compare that with the relative humidity of the air. Fog is easy to predict when you know what to look for.

Low-Ceiling Clouds

Many of the VFR-into-IMC accident victims had continued their flight as the terrain came up and the clouds came down. This squeezed them into situations where ground clearance was impossible. Could those pilots have predicted where the clouds were likely to be before takeoff? Would they have continued on if they already knew where to expect the cloud bases? We can't know the answer to those questions now, but you can use this trick to know for yourself.

As you climb, the air gets cooler; that's been established. We also know that clouds will form when the dewpoint is reached. So that means clouds can be easily predicted at the altitude where the temperature and dewpoint converge. Dry air cools off at a standard rate of 5.4 degrees F per 1000 feet. The dewpoint cools off at a standard rate of 1 degree F per 1000 feet, so the two converge at the standard rate of 4.4 degrees F (about 2.5 degrees C) per 1000 feet of altitude. To estimate the cloud bases, take the degrees of temperature/dewpoint spread and divide by the convergence rate. The answer then is multiplied by 1000 to get the cloud base estimate. Say the temperature at the surface is 55 degrees F and the dewpoint at the surface is 50 degrees. That is a 5-degree spread divided by 4.4 equals 1.136 times 1000 equals 1136 feet AGL. ($5 \times 4.4 = 1.136 \times 1000 = 1136$). A cloud layer below 1200 feet above the ground is quite low—probably not high enough for a safe flight anywhere but possibly in the traffic pattern.

Next flight, make sure to get the temperature and dewpoint from your destination and you can estimate the cloud bases with this simple calculation. Remember that the cloud bases estimate that you come up with will be an above-ground level (AGL) estimate. You must then compare that with your MSL cruising altitude. If you do this calculation and know to expect low cloud bases beforehand and then encounter low clouds while on the way, you will know that things will not improve. This will discourage you from pressing on because better weather is not likely to be just ahead.

Lives can be saved with common sense. If you are faced with deteriorating weather and don't feel completely comfortable, your good senses are probably telling you correctly to rethink your present course of action. Most pilots are not reckless. They get into trouble not because of an overconfident disregard of safety rules, but because they get caught up in the moment. Pilots rationalize that since they heard the weather briefer give a positive outlook, then the briefer must be right and the bad weather is a temporary situation, a situation I can handle. Then there is a small number of pilots who seem to get what they ask for:

NTSB Number ATL84FA054 Lexington, Tennessee

The aircraft collided with the ground in a vertical descent during IMC weather on a cross country flight. The pilot had *filed an IFR flight plan* but *he was a student pilot* with limited experience and was not instrument rated (student pilots cannot receive an instrument rating). The student was also *carrying a passenger* (student pilots are prohibited from carrying anyone except their flight instructor). The controller who had been working the flight had noted a lack of professionalism on the part of the pilot in his holding of headings and reporting procedures. When the flight finally encountered *thunderstorms* and turbulence radio contact was lost. The last radio contact was at approximately 1346. The Jackson, Tennessee FSS specialist stated to the Memphis Center controller at 1346:07 that the pilot reported that he was in severe turbulence and did not know his location. The aircraft

crashed about 4 miles east of Lexington, Tennessee. The aircraft
was a Piper Cherokee 180. Both the student pilot and his passenger were killed.

This was the most extreme case of disregard that I had ever seen. A
student pilot carrying a passenger files an illegal IFR flight plan and
flies into a thunderstorm. Thankfully, I believe that 95% of pilots
understand the responsibility of being pilot in command and want to do
the right thing. The largest killer of all, VFR flight into IMC, has the
easiest solution. Use common sense, be an instrument troubleshooter,
and add to your weather briefings by becoming your own IFR weather
forecaster.

Killing Zone Survivor Stories

The following are three short stories taken from the Aviation Safety
Reporting System, otherwise known as "NASA forms." These three
pilots lived to tell of their IMC encounters.

NASA Number: 418559

I arrived at Batten Field, Racine, Wisconsin. The weather was
poor, both there and at my destination—Pontiac, Illinois. I chose
not to leave Racine until the weather improved. I continually
called flight service, Green Bay, to look for improvement.
Weather to the south of Racine was gradually improving, but
Racine remained socked in. Later I became anxious to go in order
to get back to my business. Unfortunately, I decided to fly despite
the fog, figuring I'll climb on top and descend close to home. At
that time Pontiac had 1,900 feet ceiling. Upon climbing into the
fog, I became disoriented. I immediately called back to Racine to
let them know what was happening. I then arrested my problem
by "flying the airplane" although my body was telling me I was in
a slow, right-turning descent. Once I leveled out, I contacted Mil-
waukee departure and they vectored me through the clouds. My
decision to ignore weather in order to get back to my business
was, to say the least, irresponsible. I respected the weather, but
my business sense told me otherwise. The fact that I lived

through this experience brought home just how dangerous weather can be if one chooses to fly in it. *My trouble in flight was only a few seconds, but my memory of it will last a lifetime.*

NASA Number: 41835

Problems encountered include thickening cloud layers below the altitude I flew, and unfamiliarity with the phrase: "What are your intentions." Preflight showed no adverse weather except for broken cloud cover at southern Kentucky border for the time of arrival. Few clouds were observed 20 miles northwest of Frankfort, Kentucky, but believed them to be incomplete and would dissipate soon. Initial call-up to Lexington approach disclosed IFR and mist, plan B was to fly to Marshall Field. I was given a transponder ident and a heading to descend on. I expected to find Marshall without delay. However, the clouds were much thicker. I remember three distinct layers on my descent to 1,800 feet and thinking that my normal rate of descent would only prolong my time inside the clouds. I used carb heat, reduced power to keep RPMs below red line and increased a wings level descent to 1,800 feet per minute. I kept repeating to myself that "instruments don't lie" and remembered the hood drills and techniques I had worked on with my instructor. My awareness of the severity of my situation built slowly but now has made a permanent impression about safety issues in my mind. I overshot Marshall and had to be vectored to Cynthiana. Upon visual confirmation of Cynthiana, I signed off and thanked ATC for their help. But, instead of landing at Cynthiana, I switched my mind and opted to travel to Marshall Field by setting on a course west of the Lexington VORTAC. I found Paris (via watertower identification) and followed what I incorrectly identified as the road west to Marshall. Realizing the wrong course, I radioed Lexington Control, requested directions to Georgetown after heading back to Paris. I now more than ever realize the hazards clouds present and the importance of maintaining visual contact with the ground. I never want to go through the clouds again with so little experience. I also plan to take every effort to update my weather information from inflight advisories and metars, and an currently taking additional safety training at our FBO. I am also more determined than ever to start IFR training, hopefully next summer.

This story sounded very much like many of the fatal accident scenarios. The only difference being that this pilot was very fortunate to live to describe the flight.

NASA Number: 420767

I wanted to move my plane from its summer airport to its winter airport. The weather forecast was calling for snow the following weekend and I figured that I would not be able to get the Cessna 150 to its home base if I waited. So with 3,000 feet ceilings, I blundered off into the east skies with lowering ceilings ahead. Five minutes into the flight, I looked at the compass and said to myself that 180 degrees from this is my way out! Well, let me share the angst/terror, etc. I hit the wall of clouds and all of a sudden could not see the ground! I was only five miles away from my destination and though that I could just follow the road into my home strip! This was the biggest mistake of my life! After 4 or 5 minutes of total IMC, I finally got somewhat of a grip. I firewalled the throttle and climbed back up to 2,400 feet and prayed. I also headed in the direction that was my escape. I totally relied on the attitude indicator for help. For those of you who haven't been there, let me tell you now, that what you have learned about flying IMC is all true! Thoughts of how my attitude indicator sometimes decides to tumble zoomed through my head. A break in the clouds granted me a visual reference for a moment. Then more clouds...light...and a break about 4 miles back to home. I could see where my home base should be. Just when I could almost see home the ceiling enveloped me again and I descended to 2,000 feet MSL. The home base appeared right where it was supposed to be. At 110 mph I turned sharply and the runway was in sight! Slow to 80 mph, full flaps, idle, full rudder slip and I was on the ground! I taxied to my tie-down and said a prayer to my God that I was still alive. I do not write this as an adventure/thriller but to help others not to make my mistake! I could have easily been on the **NTSB** list of accidents that we read. I truly thought that I could end my life in muck. It is not funny or heroic by any means. I was stupid. I could go on and on about what happened to me but you get the point. Do not scud run or fly into **IMC**—period!

CHAPTER FOUR

Maneuvering Flight

ANEUVERING FLIGHT, especially in single-engine airplanes, is and has been one of the highest producers of fatal accidents. The maneuvering flight category is broad and can be confused with other accident types. In the previous chapter, several accidents occurred when pilots flew lower and lower beneath a cloud layer, eventually striking an object or the ground. These pilots collided with the airplane under control and one could conclude that they were maneuvering as the accident occurred. But I do not classify that as a maneuver accident but rather a weather-related accident. For this chapter, a maneuver accident will be considered as those accidents that take place in VFR weather conditions, and where a pilot loses control of the airplane during the performance of a flight maneuver.

From this loose definition, there seems to emerge three subcategories of maneuver accident. First, there are accidents that take place while a pilot is maneuvering close to the ground for some legitimate reason. These would include pipeline patrol, aerial application (crop dusting), banner towing, aerial photography, wildlife patrol, and law enforcement. The second category of maneuver accidents involves flight training maneuvers, and these accidents take place in the course of dual flight instruction. The third category of maneuver accidents makes up the largest group and involves low-altitude flight of a less than legitimate nature. This would include low-altitude aerobatics, buzzing, low passes, and other "stunts" and "antics."

Low-Altitude Applications

Bird spotting, powerline patrol, pollution monitoring, wildlife census, and others previously mentioned have one thing in common: A pilot is on the job and that job requires attention to be paid to something on the ground. In order to make these observations, sometimes the flight must be accomplished low to the ground. There are safe-altitude rules that prohibit flight that could be dangerously close to persons or property on the ground, but some of the jobs mentioned do require very close work, and it is possible to get a prearranged waiver of the minimum altitude rules for some low-altitude work. When pilots perform these jobs, they must be skilled at "division of attention." They must be able to safely split their attention between flying the airplane and doing the job. In each case the airplane is only incidental to the work being done. The airplane is only being used to place the observer in a vantage point where the work can be accomplished. If this observer is also the pilot, then that pilot must do two jobs at once very well. Of course, dividing or distracting a pilot's attention can be very hazardous. The pilots that do these jobs are usually not beginners.

The Code of Federal Regulations Chapter 14, Part 119 is a set of regulations that many pilots are not very familiar with. It is the section of the regulations that deal with Air Carrier and Commercial Operators. The section of Part 119 that applies to low-altitude maneuvering work is Part 119.1. This regulation outlines all the flight operations that are not considered a charter or air carrier operation under the rules of 119. In other words, 119.1 is a list of commercial operations that a person can do with a commercial pilot certificate without getting into hot water with other regulations. Pilots who want to build flight experience as a commercial pilot should consult the 119.1 list. Specifically the rule is 119.1 (e)(4) Aerial work operations including:

(i)   Crop dusting, spraying, bird chasing
(ii) Banner towing
(iii) Aerial photography and survey
(iv) Fire fighting
(v)   Use of a helicopter for construction or repair
(vi) Powerline and pipeline patrol

These are general categories of operation. It is possible to get something that is not specifically on the list approved. I once worked as a "turtle spotter!" Turtle spotting is not mentioned in 119.1, but nevertheless it was approved for a specific job. (I never saw a single turtle, but I would fly low along the North Carolina beach early in the morning and count turtle tracks. Apparently the turtles came up on the beach at night to lay eggs. The number of tracks gave the wildlife resources people an idea of the turtle population—all I know is that it was great flight time!)

Remember, the regulation is a list of what commercial pilots can do. This regulation seems to imply that low-altitude work should be done by people who have had at least enough experience to become commercial pilots.

Instructional Flight

To a flight instructor like myself, the term "maneuver" has a somewhat different meaning than the one given in accident reports. To me, maneuvers mean flight training maneuvers such as "S" turns, power-on stalls, chandelles, and the like. I was certainly interested to see what the accident rates involving these maneuvers might be. Fortunately, the news is good. The Air Safety Foundation's 1999 Nall Report of general aviation accidents and trends says, "Flight Instruction is one of the safest of general aviation activities." In 1998, flight instruction made up 22% of total flying activity, but 13.4% of total accidents. Of the fatal accidents, those involving flight instruction were 5.6%. From these numbers it is clear that flight training maneuvers are a small safety risk. The "maneuvers" that are referred to in accident reports are largely something different.

Many of the nonfatal dual flight instruction accidents were attributed to communication errors. One accident involved a student and instructor simulating an engine-out procedure. The student pilot went through initial checklists and then glided the airplane to a position where a landing in a field was possible. The instructor had intended to set up the approach, but then make a go-around. When the time to make the go-around came, there was confusion among the two as to who should execute the go-around. The delay continued until the airplane could no longer go-around due to trees at the far end of the field. The two landed and the airplane was damaged. No one was hurt, but can you imagine the argument that must have been taking place when the farmer drove up! Flight instructors are now tested on their initial flight instructor practical test to see if they can effectively transfer flight controls. The FAA-recommended method has three steps:

1. The pilot releasing the controls says to the pilot receiving the controls, "You have the flight controls."
2. The receiving pilot says, "I have the flight controls."
3. The releasing pilot confirms the switch by saying, "You now have the controls."

Using this method would remove any doubt about who is supposed to be doing what when two pilots fly together.

Nonstandard terminology has led to accidents. A flight crew for a commuter airline allowed their turboprop airliner to run off the end of a runway and collided with approach lights. The first officer was flying an instrument approach as the captain made suggestions. The airplane broke out of the clouds well above minimums, but as the approach continued the captain perceived that the airplane was too high. The captain determined that a go-around should be initiated, so he said, "Takeoff power!" The captain wanted full power just like on takeoff. But the first officer misunderstood this instruction. He thought the captain had told him to take off the power as in remove the power. The first officer chopped the power, landed long and fast, and ran off the runway. No one was hurt, but can you imagine the argument that must have been taking place when the crash trucks drove up?

Some communication accidents are attributed to equipment errors where the pilots could not communicate for some reason. Headsets and intercom systems can be very frustrating. Usually, when you can't hear each other through the intercom, or you can't hear when the other transmits, it's just aggravating, but it can also lead to problems and has caused accidents. Be careful when you rent an airplane and have to hook up an intercom. All systems are not alike and you can easily get the wrong plug in the wrong jack. Make sure to talk among yourselves and get a radio check before departure.

Low-Altitude—Personal Flying

This category of accidents is the largest of the three maneuver accident groups. In fact, 68.3% of fatal maneuvering accidents took place on what was classified as a personal flight. I would like to think that in its purest sense an "accident" is something that just happens and it is mostly beyond any person's power to prevent. Based on that definition, the "accidents" in this category are really not accidents at all. They are deliberate acts that defy safety rules, aircraft limitations, and good old common sense. I will continue to use the word accident throughout this section, but you already know how I feel about them.

Figure 4.1 illustrates all the maneuvering accidents that took place from 1983 through 2000. The accidents are plotted against the flight hours of experience the pilot had when the accident took place. These are all the accidents together, so fatal—as well as those with serious injury, and even a few with no injury—are mixed in. You can see the pattern that has been present in past evaluations of flight experience data. There is a zone where the most accidents occur. The span of experience from 50 to 150 hours is "off the chart."

These accidents largely fall into two groups: The first are low-altitude aerobatics:

NTSB Number: FTW96FA089. Vinita, Oklahoma

A witness who had just flown with the pilot said the airplane took off and remained in the traffic pattern. He thought the pilot was going to do a touch-and-go landing. When the airplane was on downwind leg and abeam the runway threshold, it banked right and departed the traffic pattern. It climbed to an estimated altitude between 2,000 and 2,500 feet MSL. Suddenly, the nose pitched up about 70 degrees above the horizon and the pilot performed a "hammerhead stall" (wing-over maneuver). The pilot recovered from the maneuver, the airplane fell off on its left wing and entered a two-turn spin to the left. The pilot appeared to recover from this maneuver also, then the airplane disappeared behind trees and impacted the ground. The witness said that on previous occasions, he had observed the pilot do this and other aerobatic maneuvers after takeoff.

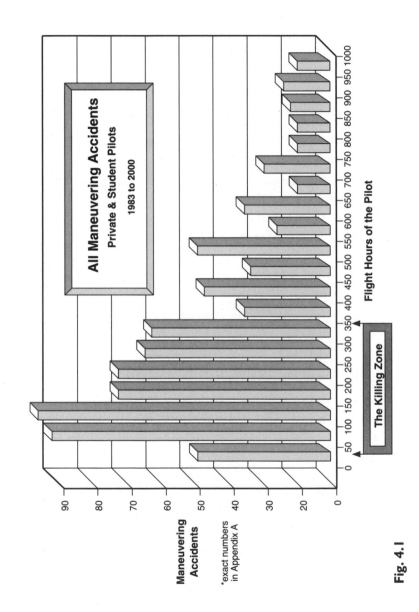

Fig. 4.1

Probable Cause:
Failure of the pilot to recover from a spin. Factors relating to the accident were: the pilot's ostentatious display and lack of altitude (terrain clearance) for recovery from aerobatic maneuver and subsequent spin.

This was a private pilot flying a Cessna 150M on a VFR day. The pilot was killed on impact. The only good thing you can say about this is that it was fortunate that the pilot let out the passenger before the stunt.

NTSB Number: LAX89LA262. Dana Point, California

Witnesses observed the airplane making slow circles, counterclockwise, when the nose of the airplane dropped and the engine sound increased. The nose of the airplane then rose to a near vertical climb and the airplane "slowly looped" over onto its back and then dove near vertically while slowing rotating 180 degrees and striking the Pacific Ocean nose first. Witnesses stated the engine continued making noise until the airplane struck the water. The aircraft was at approximately 500 feet above the water while circling and climbed to about 800 feet before stalling.

Probable Cause:
Pilot in command's failure to maintain control of the airplane while in maneuvering flight.

This was also a private pilot in a Cessna 172H, also flying alone, and also fatal.

NTSB Number: LAX89FA120B. Casa Grande, Arizona

A Private Pilot flying in his own experimental category acrobatic biplane, planned to fly in formation with a Commercial Pilot flying in his own [Cessna] normal category airplane. Both airplanes were radio equipped. While the Cessna was in cruise flight, [the experimental airplane] overtook the Cessna from the rear, maneuvered inverted, and collided with the Cessna.

Probable Cause:
The pilot in command's performance of aerobatic maneuvers, while flying formation, and his failure to maintain proper clearance.

Do you remember the scene from the movie "Top Gun" where one fighter pilot rolled inverted and took a snapshot of another airplane just below? This accident happened after that movie came out. The extra tragedy of this accident is that there were three people in the Cessna—a total of four deaths from what could easily be described as a stupid stunt.

These three are representative of this case of accidents. Aerobatic flight can be safe and fun, but like anything it is dangerous in the hands of people who don't know their own limitations or simply don't know what they are doing. If you want to fly aerobatic, seek out the services of a competent flight instructor and then use common sense.

The second groups of low-altitude maneuver accidents I call the: "fly over my house" accidents.

NTSB Number: SEA97FA178. Clackamas, Oregon

The newly certificated Private Pilot, a recent high school graduate, had two friends from his high school aboard. Before the accident, the aircraft was seen maneuvering at low airspeed and altitude in the vicinity of one of the passengers' houses. A flight instructor-qualified witness reported that at about 40 feet above ground level, the aircraft entered a steep turn of more than 80 degrees bank, and that the nose then dropped steeply and the aircraft crashed. The crash site was $\frac{3}{8}$ mile east of the front seat passenger's house. This passenger's brother reported that he was talking with the front seat passenger on a cell phone at the time of the accident. He stated that the passenger instructed his brother to come out on the deck of the house because they were going to fly over. When the brother told the passenger he had not seen the aircraft, the passenger told his brother to wait while the aircraft turned around and came back over. During an on-scene examination of the aircraft wreckage, investigators found no evidence of preimpact aircraft mechanical problems.

Probable Cause:

The pilot's failure to maintain airspeed in a steep turn and his concurrent failure to maintain adequate altitude, resulting in an accelerated stall at an altitude insufficient for recovery. Factors included the pilot's intentional low altitude flight and maneuvering in a ostentatious display, and the pilot's lack of total experience.

Not mentioned in the NTSB's probable cause was a certain amount of peer pressure that the pilot was obviously not ready to handle. This was a Cessna 172M. The "newly" certificated private pilot had 87 total flight hours. One passenger survived the crash with serious injuries; the pilot and the passenger with the cell phone were killed.

NTSB Number: ATL88FA078. Cedartown, Georgia

Ground witnesses reported that the aircraft flew repeated circles at treetop height around the residence of the pilot's parents. While in a steep bank the aircraft pitched nose down and descended until impacting the ground. The pilot in command had been issued a student pilot certificate but was not currently endorsed to solo. The aircraft was flown without authorization from the owner/instructor. Wreckage exam showed that the stall warning was inoperative. The owners manual includes a check of the stall warning system during preflight inspection. 14 CFR 43 requires certification of the system during annual inspection of the aircraft.

As stated in the report, this was a student pilot who essentially stole an airplane. The pilot had 58 total flight hours and died in the crash.

There are several other accidents that have this same format. The crash site is within close proximity to a girlfriend's house, a boyfriend's house, a school, or a parent's home. I was 17 when I got my private pilot certificate, the minimum age, and I can remember wanting everybody at my high school to know about it. I understand the temptation. All the "cool" kids were able to show their talents in front of an audience: playing football, cheerleading, even drama. I believe that flying an airplane is cooler than all that, but nobody was there at the airport when I "performed." I am surprised there are not more buzzing accidents. Young pilots, let me recommend that you impress your friends by inviting them on a flight to eat dinner somewhere, or to go to a ball game. The pilots in the accidents of this type probably had good student instruction and knew better. Part of becoming a pilot is accepting the responsibility of being pilot in command. That responsibility will call for you to say NO! more often than you might think.

Simply stated: Stunts and antics are not professional. Each of us as pilots should want to accomplish every flight with the greatest safety. We will not always turn in a perfect performance, but we should always desire and work toward the picture-perfect flight. When I hear of someone who does make a dangerous maneuver, I always think that the person at the wheel is like a modern-day Icarus. He or she has been given the gift of flight but has not learned how to handle it. Mythology tells that the foolhardy Icarus ignored the advice of his wise father, flew too close to the sun, and crashed to earth. Low-altitude maneuver accidents could end forever if pilots would fly like professionals.

Aerodynamics of the Turn

Avoiding low-altitude steep turns and low aerobatics all together is the safest course of action. But what is it about these maneuvers that become deadly at low altitude? Why does the airplane seem to fall out from under a pilot when in these maneuvers?

The airplane's wings must provide lift to counteract all "down" forces. Weight or gravity is the "down" force that we easily understand, but while flying other forces come into play. These additional forces can team up with gravity and reduce the effectiveness of lift. Figure 4.2 illustrates two airplanes in flight. The airplane on the left is flying straight and level. The lift exactly opposes weight. These lift and weight vectors are fairly simple, but things get complicated when the airplane turns. The airplane on the right is in a medium bank turn. The first problem is that the lift vector is now leaned over, in the turn. Between the two airplane diagrams is a comparison of the "effective" lift. You can see that when the lift vector is leaned over, we lose effective lift because the lift vector no longer directly opposes weight. So in a turn we lose lift.

Meanwhile, the turn will also produce centrifugal force. This is the swaying force you feel in your car when you take a fast turn. Centrifu-

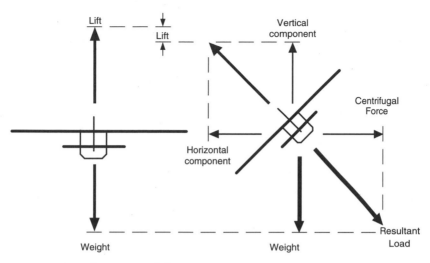

Fig. 4.2 Aerodynamics of the turn.

gal force joins forces with gravity to form a resultant load. This is more commonly called the G force. The actual force of the earth's gravity does not get stronger when you turn, but when you add gravity and centrifugal force together it places an extra load on the wings. From the wing's point of view, it is being asked to carry a greater load.

The wing is being asked to carry a greater load at the exact moment when lift is reduced and the wing is less able to carry a greater load. Something has to give. The accelerated stall takes place. Ordinarily the stall speeds are painted on the airspeed indicator. The slow end of the white arc is the stall speed with flaps down and the slow end of the green arc is the stall speed with flaps up. But in a turn the colors of the airspeed indicator can no longer be trusted. The airplane can and will stall even though the airspeed is well within the green arc. It stalls faster than the indicator says it should; that's why its called an accelerated stall.

Figure 4.3 is a chart of load factors. You can see that at shallow banks, the G force is not much above 1G. But when a pilot makes a 60-degree level turn, the G force jumps to 2Gs. That means a 2000 pound airplane now effectively weighs 4000 pounds in the turn. More importantly, the wings must support 4000 pounds. That is a great deal to ask—to get 4000 pounds of lift from the wings of a 2000-pound airplane. The wings probably will not be able to do it and lift is lost; the airplane stalls.

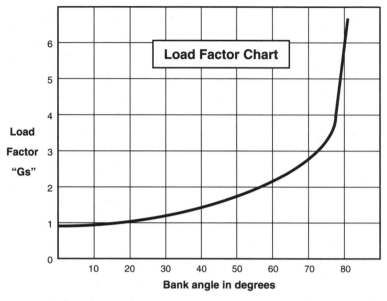

Fig. 4.3 The airplane's wings must support more effective weight as the level bank angle increases.

NTSB Number: CHI95LAI65. Osage Beach, Missouri

Witnesses observed the airplane liftoff at the runway midpoint. One witness observed the airplane begin a turn to the right when it was about 50 feet above the ground. He said the airplane continued to bank to the right and descend into trees, erupted into flames shortly after colliding with the ground. The pilot said he was between 500 and 800 feet above ground when he began a right turn to the departure runway's downwind leg. He said the airplane slipped more to the right than he intended. He said it had slipped into about a 45-degree bank and that the stall horn activated. The pilot said the airplane would not level off with left rudder application. He said he had not used left aileron to help with the recovery. During the on-scene investigation, no mechanical anomalies were found with the airframe or engine.

This was a private pilot with 327 hours who survived the accident but had serious injuries. There was one passenger on board, and that passenger was killed. There was a big difference in the altitude estimates of the witness and the pilot: 50 feet versus 500 to 800 feet. There was no mistaking what happened: "a 45-degree bank and the stall horn activated"—accelerated stall.

The accelerated stall phenomenon is also why it is best to land straight ahead if an engine fails after takeoff. The terrain just ahead might not look too good—trees, houses, etc. But it is much better to land straight ahead with the wings level than to attempt a turn and risk an accelerated stall while in the turn. If an accelerated stall does take place at low altitude, there will be virtually no chance for recovery. Remember the accident where the passenger was talking on the cell phone? A flight instructor on the ground who was a witness said that the airplane attempted a turn of 80-degrees bank! Refer to Figure 4.3. An 80-degree bank could produce 6Gs. That resulted in a stall that was completely unrecoverable and fatal.

If you want to learn to fly aerobatics there is a right way to do it. Seek out the services of a qualified aerobatic instructor. Aerobatics instruction is optional now, but many believe that the basics of aerobatics should be required at least for advanced certificates. Could exposure to aerobatics assist pilots who find themselves unexpectedly out of control to regain control? One aerobatics instructor I know advocates minimum aerobatic instruction for the ATP. He cites some past airline accidents where control was lost. Could the pilots have managed to regain control if they were more familiar with aerobatics? We will never know for sure.

The numbers show that most of the maneuver accidents could have been prevented if a little flight discipline had been used. We can look at the aerodynamics of why these stunts are dangerous, but the mature pilot doesn't need physics to know that low-altitude maneuvers are deadly and unprofessional.

CHAPTER FIVE

Takeoff and Climb

THE SAYING IS THAT "landings are mandatory; takeoffs are optional." This means that we can always say no to a takeoff if we choose, but once airborne, we have no choice but to make a landing somewhere. The accident pattern for takeoff and climb looks much like other accident categories. The fatal takeoff accidents again have a peak of accidents that is contained by another killing zone from 50 to 350 flight hours experience of the pilot (Figure 5.1). The exception is that student pilots, although having a small number of fatal accidents (see Figure 5.1 at the student pilot range: 0 to 50 hours), have the most non-fatal takeoff accidents. The pilots with between 0 and 50 flight hours had the greatest number of takeoff accidents, with 293 from 1983 to 2000. These included the 10 fatal accidents (Figure 5.1), but most were accidents that involved no injury at all. This implies that takeoffs are hard to learn since more beginners have accidents.

During a two-hour flight, how much of the time is spent completing a takeoff? That would, of course, depend on the flight. This is where general aviation has a much greater exposure to takeoff risk than other areas of aviation. If the two-hour flight is a cross country, taking off once at the departure airport and landing at the destination, the takeoff may only be 2% of the flight. But if it is a flight instructor and student working on takeoff and landings, the takeoff portion may be 50% of the flight. Since the exposure to risk varies widely, it is hard to say that takeoffs are more dangerous than another phase of flight. But looking at only fatal accidents among student and private pilots since

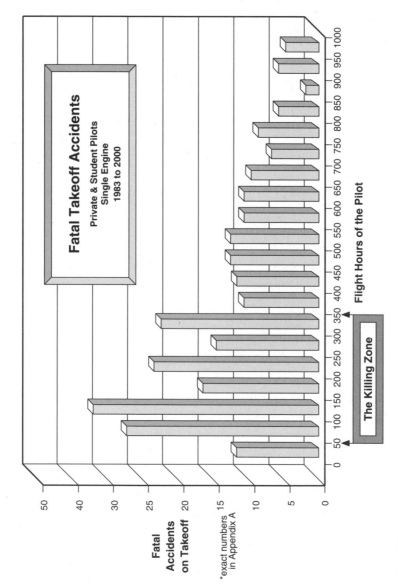

Fig. 5.1

1983, approximately 10% have taken place during or immediately after takeoff.

The takeoff accidents themselves fall into three identifiable subcategories: improper airplane configuration, poor pilot planning, and engine failure.

Not Ready for Takeoff

There are an alarming number of accidents on takeoff that happen because the pilot does not get the airplane ready but attempts to take off anyway. These are cases of improper airplane configuration, some are due to haste, and others are caused by a lack of knowledge.

NTSB Number: FTW93FA068. Porter, Texas

During the initial climb following takeoff the airplane stalled and descended approximately 60 degrees nose down approximately 440 yards from the departure end of the runway. Witnesses reported the airplane attempting to take off with more flaps extended than required by the "normal" procedures. The airplane flight manual states that flap deflection of 30 and 40 degrees are not recommended at any time for takeoff. The flap indicator in the airplane was found at 45 degrees, while the flap actuator was measured as 38 degrees of extension. No anomalies were found in the flap operating or indicating system. An inspection of the engine also revealed no anomalies. This was the pilot's first flight in this type of airplane without an instructor onboard.

Probable Cause:

Inadvertent stall. Factors were the pilot's failure to follow published checklist and the improper flap setting.

It is not clear whether the pilot may have run the flaps down during the pretakeoff checklist, become distracted, and failed to raise the flaps, or was the pilot deliberately making the takeoff with flaps down? Either way the flaps were down and the airplane could not climb against the drag produced by that degree of flap extension. The pilot was a private pilot with 138 hours flying a Cessna 150J with one passenger. Both the pilot and passenger were killed.

NTSB Number: MIA95FA200. Pulaski, Tennessee

The pilot did not perform an engine run-up before takeoff. During the takeoff roll, all the available runway was not used and witnesses reported "sluggish acceleration." The airplane rotated about 500 feet from the departure end of the runway and climbed to about 75–100 feet above ground level with a "very high angle of attack." The airplane then rolled to the right, collided with trees, the roof of a barn, the ground, and came to rest inverted. Examination of the engine revealed that the carburetor heat was partially engaged and the control cable was found disconnected from the control arm at the carburetor heat air box assembly. The castellated nut was not safety wired. The engine was installed about one month before the accident. The takeoff ground roll was calculated to be about 1,100 feet.

Probable Cause:

Failure of the pilot to perform the before-takeoff checklist, and his failure to abort the takeoff due to inadequate acceleration. Contributing to the accident was an inadequate annual inspection by maintenance personnel for failure to safety the carburetor heat control cable, which resulted in the disconnection of the carburetor heat control cable from the control arm and inadvertent partial activation of carburetor heat.

This Piper Cherokee 140, like most small airplanes, uses a heat transfer from the engine exhaust to provide carburetor heat (Chapter 12). The heat is used to melt ice from a carburetor should ice accumulate and threaten to choke off the airflow to the engine, which would stop the engine. The problem is that heated air is less dense and the result is less engine power. When air goes into the engine's cylinders, it is squeezed by the rising piston and then burned with the fuel. The more molecules of air that get into the cylinder, the more power will be produced. But with carburetor heat on, this pilot was attempting a takeoff with partial power. The checklist provides an investigation into the operation of carburetor heat. During the engine run-up the carburetor heat is pulled on, allowing heated air into the engine. Since the heated air is less dense, less power is produced and the pilot sees this as a drop in RPM. The pilot involved in this accident did not perform the pretakeoff checklist, so he did not give himself the opportunity to detect the problem. It did not help that he also made an intersection takeoff. It is not "illegal" to use only part of the runway, but runway behind you is useless.

This case is also a good look at the responsibility of the pilot in command when it comes to maintenance errors. We pilots are very vulnerable. On the one hand, most pilots are not also A&P mechanics and therefore we fly airplanes that other people have worked on. I usually am not present when the work is done, so I do not oversee the work. Yet I am completely responsible for the safety of the airplane, including its maintenance, once I elect to make a takeoff. This does not take the maintenance technicians off the hook; in this case part of the accident cause was an error by an A&P, but the pilot had the opportunity to detect the problem and did not do so. From the evidence it appears that the pilot was in a hurry: no engine run-up and no taxi to the end of the full runway length. An error chain is clear. A maintenance mistake and a pilot in a hurry combined to cause an accident and the loss of two lives.

NTSB Number: DEN90FA161. Hettinger, North Dakota

After takeoff at about 200 AGL, the aircraft was observed in a left wing down, exaggerated nose high attitude, and the roll rapidly to the right and impact the ground in a near vertical nose down descent. The pilot's log book indicated a total pilot time of 113 hours over a 45-year time period, with 2 hours during the last 90 days. The aircraft was approximately 74 pounds over the maximum gross weight.

Probable Cause:

The pilot's failure to maintain airspeed during climbout. Contributing factors were: the pilot's lack of recent experience, and the over maximum gross weight condition of the aircraft.

I believe an overweight takeoff attempt fits into this category because even though all control surfaces were properly set, the airplane was still not configured properly because it was just too heavy. This was a Cessna 150L flown with one passenger.

Density Altitude

Density altitude is simply a measurement of the "quality" of the air. When airplanes fly through dense air three things happen:

1. The wings produce more lift because there are more air molecules to produce lift with.
2. The propeller produces more thrust because there are more air molecules to cut through and push back.
3. The engine produces more power because more air molecules are compressed and then expanded on the power stroke.

Low altitude, low temperature, and low humidity produce dense, wonderful air for airplanes to operate in. But the converse is true. High altitude, high temperature, and high humidity make wings sluggish, props dull, and engines weak. If you are planning a takeoff from a sea-level airport on a hot day, the density altitude might be 2000 feet. In other words, the airplane will perform like it was already 2000 feet up, even though you are actually at sea level. In this situation the pilot should anticipate worse than normal performance. Many AWOS weather transmitters that are now located at uncontrolled airports will report the density altitude. Use that number, or calculate the density altitude yourself, and factor in the density altitude performance for each takeoff.

NTSB Number: ATL89FA182. Ridgeland, South Carolina

Two pilots were seated at the controls. The right front seat pilot had rented the aircraft [Cessna 172G]; he had 59 hours total flight time, of which 3 hours was in the Cessna 172. The left front seat pilot (normally the position of the pilot in command) had 120 hours of flight time, of which 52 hours was in a Cessna 172. Before renting the aircraft, the right front seat pilot was told of the aircraft's limited payload and runway length requirements due to the auxiliary fuel capacity. The aircraft had been refueled to its capacity at the en route airport. During takeoff from the 3,100 foot runway, the aircraft lifted off, settled back onto the runway and lifted again. The aircraft climbed with a steep nose high attitude to about 200 to 250 feet, then it entered a steep descent and crashed. The wind was reported to be aligned with the runway at 10 knots and the temperature was 90 degrees. The aircraft was estimated to have been loaded about 120 pounds over its maximum weight limit. No preimpact part failure or malfunction of the aircraft was found. Density altitude was approximately 2100 feet.

Probable Cause:

Improper planning/decision by the pilot in command (PIC), which resulted in an inadvertent stall during takeoff. Factors relating to the accident were: the excessive gross weight and improper lift-off by the PIC, the pilot's/passenger's improper crew coordination regarding the specific instructions that he received concerning the payload limitations, and the high density altitude.

You don't think of South Carolina as a place where density altitude would be a problem, but that is why this accident was selected. South Carolina is a low-elevation state, but hot weather can turn near sea-level airports into hazards. This accident took place after a chain of events. The pilots filled the airplane with too much fuel. The runway was short. The liftoff was premature, and the density altitude was 2100 feet. Any one of these factors taken alone would not have caused the accident, but together a crash was inevitable unless a thinking pilot were to step in.

NTSB Number: LAX94FA272. South Lake Tahoe, California

The pilot began his takeoff from the beginning of the 8,544 foot runway. An air traffic controller observed the airplane [Piper Cherokee 180] become airborne after rolling between 2,500 and 3,000 feet. Witnesses reported the airplane pitched up and down several times and was flying slowly as it climbed between 100 and 200 feet AGL. While still over the runway, and out of ground effect, the airplane commenced a steep (45 to 90 degree) left bank and crashed into a field 900 feet east of the runway. Wreckage and ground impact signatures were consistent with the aircraft having collided with the terrain while in an 80 degree left bank. No mechanical malfunctions were found with the airplane. During the preceding 20-month long period since the pilot received the Private Pilot Certificate, he had flown airplanes for 6 hours. His total experience in the Piper airplane was 3.1 hours of which 1.4 hours had been a checkout flight given by a CFI. The CFI reported he had not checked out the pilot at a high-density altitude airport. The calculated density altitude was about 8,570 feet.

Probable Cause:

The pilot's failure to maintain adequate airspeed during initial climb under high-density altitude weather conditions and a resulting inadvertent stall/spin. Factors which contributed to the accident were the pilot's overconfidence in his personal ability, and his lack of experience flying the airplane.

Overconfidence combined with a lack of experience is a deadly mix. This airplane was not overweight, but the high elevation of Lake Tahoe and the hot July afternoon temperatures joined forces. The wings, propeller, and engine were gasping for air and could not do what the pilot asked them to do. Even so, there was still runway ahead when the pilot got too slow and stalled. The pilot was so unaware of the performance problem present that day that even with an opportunity to put the airplane back down straight ahead, he still pushed the airplane to fly when it could not. The fact that the airplane got airborne at all was due to ground effect, as eluded to in the report. When the wings produce lift, air is accelerated over the airfoil. When this accelerated air reaches the wing's trailing edge, it overpowers the slower air that joins up from under the wing and creates a "downwash." The air behind and under the wing is pushed down. In normal flight this does not matter, but close to the ground the downwash quickly strikes the surface and bounces back up into the wing. This provides a "cushion" of air that the airplane can ride on. It is most pronounced in low-wing airplanes because obviously the underside of the wing is closer to the surface and this squeezes the air between the wing and the ground. This ground effect can fool a pilot. The airplane can leave the ground, but that does not mean that it can climb away from the ground. In a high-density altitude condition, it may not be possible for an airplane to climb above ground effect, but the pilot who becomes airborne may think that everything is a-okay because the airplane first came off the ground normally. This is a real pilot trap.

NTSB Number: NYC91FA140. Shelburne, Vermont

The airplane took off from a 2,500 foot grass strip with three adults on board. According to witnesses and one of the passengers, the airplane lifted off, then about halfway down the runway, it settled back to the surface. The witnesses said that the airplane got near the end of the runway, lifted off again and started a slow climb. The passenger sitting in the right front seat, "we took off but the aircraft climbed very slowly and never really climbed out very high...we were in a slight right turn when we hit the trees..." The winds were light, but the passenger said, "...the direction of the wind was with our direction of flight..." About 15 miles north at Burlington, Vermont, the wind was from 280 degrees at 6 knots. Performance charts showed that (for a paved runway), 2,600 feet would have been required to clear a 50 foot obstacle.

Probable Cause:

Premature lift-off by the pilot, which resulted in a subsequent stall/mush and collision with trees. Factors related to the accident were: the pilot's inadequate preflight planning/preparation, and the unfavorable wind condition.

Some runways, especially grass runways, will have a slope to them. The slope may be so steep that it becomes more of a concern than the wind. In these cases the best way is to land uphill and take off downhill regardless of wind direction, but watch out! You cannot completely ignore the wind. Putting all factors together, a downwind, downhill takeoff might require more runway than is available. In the last accident example, the airplane manual, which presumably went down in the accident, held numbers that clearly told the pilot that more room was needed (2600 feet) than was available (2500 feet) to clear the trees. The answer was in black and white, but ignored.

NTSB Number: ATL89FA081. Hartsville, Tennessee

Takeoff was from a private strip in a cow pasture. The first half of the strip was on the crown of a hill. It sloped down past the midpoint of the runway. The passenger in the rear seat survived and reported that during takeoff the airplane hit a little ditch,

became airborne, turned left, and climbed with a nose high attitude. The engine "did not miss a lick," he said. The pilot's wife
watched the takeoff and said when the airplane lifted off the
ground it started a left bank and then collided with a tree about
300 yards from the lift-off point. After the collision the airplane
continued to turn left and hit the ground nose first, vertically.
Examination of the aircraft and engine did not reveal any preimpact malfunctions that would have affected the flight.

Probable Cause:

The pilot allowing the aircraft to lift off prematurely at a slow
airspeed and his failure to maintain directional control before
inadvertently entering a stall. Contributing to the accident was
the pilot's experience level and lack of qualifications.

That last reference to "lack of qualifications" was included in the
probable cause statement because this pilot was only a student pilot!
That's right, a student pilot with 33 hours carrying two passengers.
Student pilots are absolutely prohibited from carrying passengers.
Student pilots also have a very good safety record as we have seen
through the book, but this accident happened at "a private strip in
a cow pasture" and probably away from the watchful eye of a flight
instructor.

The last two accident examples had another thing in common.
They both involved a "premature liftoff." Why is a premature liftoff
such a problem? If the pilot rotates for takeoff at a speed that is too
slow to produce enough lift, the airplane will stall or mush. The airplane may become airborne at this slow speed due to the ground effect,
but it will not safely climb. Why would a pilot attempt a rotation too
soon? Most premature liftoff accidents take place on short runways.
The downwind accident in Vermont was on a short runway. As the
takeoff roll continues, the pilot sees those trees at the far end getting
closer and closer. They pull the airplane off too soon, because they are
uncomfortable with those trees staring them in the face. But a premature liftoff will ultimately prolong and extend the takeoff roll closer to
the trees. The airplane can't climb, so it "settles" back to the runway,
but it is going no faster than before. Now the trees really look close
and the pilot will be tempted to pull back again. This is a dangerous

cycle that can't end well. The solution: plan the takeoff before you see the trees accelerating towards you—before you even get on the runway to begin with. Using the proper liftoff speed, determine that this liftoff speed can be obtained within the available runway length. Armed with this information, make the takeoff and stare down the trees. It takes cool patience to allow the airplane to accelerate down a short runway, the trees at the far end looming larger, but until the proper liftoff speed is obtained you are not going anywhere. When pilots get nervous feet and pull the airplane off early, they actually make the situation worse. The best way to attack this problem is to do the math. Know in advance about the runway length, wind, density altitude, and your airplane's specific performance. With this knowledge you will not have nervous feet.

Engine Failure

An engine failure on takeoff may be the pilot's greatest challenge. The pilot's immediate reaction will be the difference between life and death. If you get airborne over the runway and the engine gives you any trouble, your best solution is to reduce power and land straight ahead on the remaining runway. This is why we leave retractable landing gear down while there is any runway or clear zone ahead. The toughest problem exists between a point after the runway has passed behind and before enough altitude has been gained to turn around to the runway. More altitude will be lost in a gliding turn-around than in a straight-ahead glide. This fact makes the 180-degree turn at low altitude very hazardous. In fact, such a turn would actually be greater than 180 degrees. The turn back to the same runway as the takeoff would be more of a teardrop shape with an initial 210-degree turn followed by at least a 30-degree turn in the opposite direction to line up with the runway. The attempt to turn around and get back to the runway after an immediate engine failure has been termed the "impossible turn." Review the "Aerodynamic of a Turn" in Chapter 4 and you will understand that a turn places additional load on the airplane and raises the stall speed. In the instant that an engine failure occurs on takeoff, the pilot must remember that a turn back to the runway from a low altitude may be physically impossible.

NTSB Number: MIA97FA014. Smithville, Tennessee

The flight departed on runway 24 and shortly after crossing the end of the runway, the engine suddenly lost power. The pilot transmitted on the unicom frequency that he had a problem and was turning back. The aircraft was observed at 200 feet above ground level in a steep left bank. After reaching a northeasterly heading, the nose and left wing of the aircraft dropped, and the aircraft descended nose first into a wooded area where it crashed. A post crash fire erupted and consumed the fuselage and left wing.

Probable Cause:

Loss of engine power for undetermined reason(s), shortly after takeoff, and failure of the pilot to maintain adequate airspeed, while turning in an attempt to maneuver back to the airport, which resulted in an inadvertent stall and uncontrolled descent into wooded terrain. The pilot's attempt to return to the departure airport was a probable factor.

NTSB Number MIA00LA127. Murfreesboro, Tennessee

An instructional flight crashed after takeoff in the vicinity of Murfreesboro, Tennessee. Visual meteorological conditions prevailed and a VFR flight plan filed but not yet activated. The airplane was destroyed by the post flight fire, but the CFI-rated pilot and student pilot were not injured. The flight had departed about three minutes before the accident. According to the airport manager, the flight had departed runway 36 at Murfreesboro Municipal Airport, sustained a loss of engine power, and collided with powerlines on the airport's perimeter. The empennage separated and remained entangled in the powerlines. The fuselage landed at the base of trees and burned.

These two accidents were very similar. Both were single-engine airplanes on takeoff from uncontrolled airports. Both experienced a yet-unexplained loss of engine power. That is where the similarities stopped. The pilot of the first accident attempted to turn back to the runway, but in order to get turned all the way around in enough time, a steep bank was needed. The steep bank, however, produced an accelerated stall and the pilot lost airplane control. In the second accident, the flight instructor at the controls did not attempt a 180-degree turn.

Figures 5.2 and 5.3 are of this second accident. Look closely at Figure 5.2 and you will see the airplane's empennage caught in the powerlines and the trees above. The fuselage, engine, and wings can be better seen in Figure 5.3. You can see one blade of the propeller, but the cockpit was completely turned to ash. The powerline and trees stopped the airplane's forward progress. The tail snapped off and the student and instructor fell to earth upside down. The instructor's side door would not open, so both came out the left side as the fire ignited.

I spoke to this CFI after the accident. He said that when the power loss took place, he first thought about turning around but quickly discarded the idea. The view ahead could not have been very good. He knew that he would not clear the row of trees and the powerline. He also knew that there was 4000 feet of clear runway just behind him. He verbally switched airplane control from the student to himself. It must have

Fig. 5.2 The airplane lost power after takeoff, but the pilot flew straight ahead. The airplane hit trees and a power line. The tail is still caught in the power line.

Fig. 5.3 The cockpit, engine, and wings fell to the ground and burned. The flight instructor and student escaped unhurt.

been very tempting, but he kept his head and he flew the airplane. He did not allow the airplane to stall and he flew under control into the trees. A controlled flight into the trees was better than an out-of-control plunge. The pilot that attempted the "impossible turn" was killed, the instructor and student who did not turn back were unhurt.

Prevention

The good news is that most takeoff accidents could have been prevented. We are not powerless to stop and reverse the trend of takeoff accidents. Some takeoffs cannot physically be made safely, and the information that tells the pilot this is contained in the pilots' operating handbook or information manual. Do not fly an airplane until you have looked over its takeoff performance charts. What if you were driving down a country road and came upon a roadblock that said, "Danger—Bridge Out." But instead of turning around, you threw the sign in your back seat and drove around the roadblock. Ignoring the warning that was in your possession, you speed over the bridge, find yourself temporarily airborne, and then hit the water below. Aren't pilots who take off when their airplane manuals warn them against it doing the same thing? It is very frustrating for accident investigators to find airplane manuals in the wreckage that contain the warning to the pilot that could have prevented the accident in the first place. Understand what the effects of density altitude, wind, runway surface, and runway slope will have on your airplane's takeoff performance. Many of these accidents took place in rental airplanes. This might mean that the pilots were less familiar with the airplane than had it been their own, but this is no excuse. The reason that airplane flight manuals and pilot's operating handbooks are required for flight is so that every pilot will have access to information regarding performance.

Many accidents have been caused by improperly configuring the airplane before takeoff, and in those cases a pilot in a hurry was a factor. The best solution therefore is to slow down. This, I know, is easier said than done. We have all found ourselves behind schedule and pressed to get into the air and under way. I am guilty of this myself. But I always try to remember that there is nothing more important than preparing for flight. I might rush through lunch, but not pretakeoff inspections.

Sometimes the situation prior to takeoff can place you in a rush. There will be the day when you are meticulously completing your pre-takeoff checks and you hear the purposeful revving of an engine just behind you. It's another pilot who is telling you that he is ready to go and you are in the way. At large airports it's not uncommon to look up from your checklist to see the nose gear of an airliner filling your window. Air traffic controllers will also pressure you sometimes, "N4321A about how much more time will you need out there?" All this can easily pressure a pilot into accepting a takeoff with the checklist partially completed. To avoid being in this situation, try to use the taxiway's expanded run-up areas where available. Taxi as far off to the side as you can, even when you can't see anybody behind you yet. Turbine airplanes often will not stop for a run-up, so you will often hear them tell the tower, "We'll be ready at the end." Translated, that means that they are already prepared for takeoff, they are late, and they hope ground control didn't put a "little guy" in the way. First-come/first-serve really does not apply at the end of a runway. I am always happy to let a faster airplane go ahead because I really don't want them taking off behind me anyway. If you are ever pressured to go before you are ready by either another pilot, ATC, or even your own passengers, remember who is pilot in command. Sometimes you'll have no choice but to block the taxiway. If another airplane comes up and is ready to go while you still have several minutes of pretakeoff preparation to complete, it would be better to taxi out onto the runway (with permission at controlled airports, of course) and exit again at the first taxiway. This makes the pilot behind you number one for takeoff and buys you the time you need to be safe.

I use my mobile phone all the time now to pick up IFR clearances from uncontrolled airports. Using the mobile phone in flight is prohibited by the Federal Communications Commission (FCC), but there is no problem while taxiing out to the runway. Before mobile phones, you had to get a "void time" using the FBO telephone or a phone booth. The void time allowed an IFR pilot to take off into IFR conditions and climb to an altitude that was high enough to establish two-way communications with ATC. The problem was that controllers could not (or would not) allocate the airspace to one airplane for very long. Often the void time was only a 10-minute window. This forced the pilot to leave the phone, leap into the airplane, taxi out, run-up, and be in the air within that 10 minutes. A short void time always created a scramble to get into the air; there was a flurry of hurry accidents under void time restrictions. Today that is not necessary. Just board passengers, taxi, and run-up at a safer pace. When you are completely ready for takeoff, then call for the clearance. In your local area you can get the telephone number of the clearance delivery nearest your departure for the Airport/Facility Directory. If you do not have the number, then call 1-800-WX-BRIEF, which is the nationwide Flight Service Station (FSS) telephone number. When FSS answers, they will not have your clearance right there, but they can call ATC to get it. When they ask, "How much time will you need?" I love to say, "I'm ready right now and I'm coming off runway 18." Using the mobile phone eliminates the rush and keeps the pilot in control.

Another group of takeoff accidents happened when the pilot attempted to climb too early before a safe climb speed was achieved. We saw that this happens on short runways, when pilots get nervous that they are running out of room, and rather than watching the airspeed they are watching the distant trees. There are three problems with a premature liftoff. First the airplane stalls off the ground into ground effect but will not climb much higher. When the airplane comes off the ground the pilot expects that it will continue climbing as normal and has a hard time believing otherwise when the airplane mushes. "What is happening here?" the pilot thinks and pulls back more on the yoke. Many of the accident examples had witnesses that said that the airplane's nose went up to an exaggeratedly high pitch before pitching down. This happened because the pilots were still under the false impression that there was something that they could do about the situation. They painted themselves into the corner, but still could not accept the fact that they were stuck. The second problem is that when the noise is pitched up, the airspeed will decrease. It is only an increase in airspeed that will remedy the situation, but that requires the nose to be down. If the pilot had allowed the airplane to completely accelerate to a safe liftoff speed, even in the face of obstructions ahead, the total takeoff roll would be less than a pilot who pulls off too soon, does not climb, does not accelerate, and settles back to the runway. In that second case, if there is still enough runway remaining to accelerate to takeoff speed, the airplane may still be able to make a normal takeoff, but look at all the runway that was used up. A premature liftoff will increase the ground roll in an attempt to reduce the ground roll! If the pilot knows the takeoff distance ahead of time by referring to the airplane's takeoff charts, then there will be less surprise on takeoff and less temptation to lift off early. The last danger from an early liftoff comes from any crosswind that might be present. When the airplane comes off the ground with a slow speed, it becomes a sitting duck for the cross wind. There will not be enough vertical space to do much wind correcting when you are only at the altitude of ground effect. When the airplane is slow, any wind will have a greater effect on the airplane. Who needs a greater wind correction angle to overcome an en route crosswind, a Lear Jet or a Cessna 150? Even a light wind can blow a slow-moving airplane off the centerline when the airplane is in ground effect. If the airplane starts to be pushed off the runway, the trees that line the runway will be much closer than the trees at the end of the runway. A premature liftoff during crosswind conditions means the pilot cannot climb, cannot accelerate, and cannot prevent the wind from blowing the airplane into obstructions downwind. A pilot in that position becomes a passenger who can do nothing more than ride it out. No pilot wants to be reduced to having the same control as a passenger, so plan ahead, read the charts, take your time, and be ready to say no.

Short-Field Takeoffs

A short-field takeoff is a maneuver that is required of student pilots to learn, but not for the reason you would think. If an airplane were actually caught out at a short field where a safe climbout was in doubt, do you think we should get a student pilot to fly it out? Probably not. Instead we would get someone with a little more experience. So why do we teach short-field takeoffs to student pilots if we would discourage their use of the maneuver? It is the elements of the maneuver that are important. We teach student pilots to control airspeed accurately in a critical situation. During a short-field takeoff, the airplane is deliberately flown with just a few knots of airspeed above a stall, and just a few feet above the ground—that's critical! If the student can nevertheless safely handle that critical situation, the FAA called that "meeting the standards." If airspeed can be controlled, then there can be no stall and therefore no stall/spin takeoff accident.

The takeoff should not be attempted without taking advantage of all available runways. In training we usually simulate this by taxiing as close to the end of the runway as possible, then turning to line up with the runway centerline. If a pilot faced a real (not simulated) short field with real danger from obstructions, they could do more. The pilot could stop the engine and pull the airplane backwards so that the main wheels are at the threshold of the available runway. The load would be reduced. Any extra weight would be removed—even extra seats (if done by an A&P and the weight & balance forms adjusted). The takeoff could be made with reduced fuel, leaving enough to fly to the next airport with ample runway length.

The technique for a short-field takeoff is to line up with the runway and then add full power while holding the brakes. Holding the brakes does not shorten the ground roll. The reason the brakes are held is to give us time to inspect the operation of the engine. The pilots must make sure that they are developing full power and that engine instruments are in the green before committing to the takeoff. If the engine were not developing full power, you would rather not discover this halfway down the runway or just after liftoff. Having full power when the brakes are released as opposed to a gradual increase of power while the airplane is rolling will reduce the ground run.

Before a short-field takeoff, do a thorough run-up inspection of the engine. Look for anything that could reduce power output like fouled spark plugs, improperly timed magnetos, or carburetor heat valve problems. Use the airplane's checklist to check for proper magneto drops, and difference between magneto drops. In addition to your ear, you can detect a rough engine with your feet. While holding brakes, you can feel small vibration changes through the rudder pedals. Only after you are sure the engine is giving you all it has to give should you release the brakes to begin the takeoff roll. When brakes are released, torque will be high, so be ready to apply right rudder if the airplane comes out of the brakes veering to the left of the centerline.

The initial acceleration always seems very slow when trees are ahead. But if you have done your homework—you understand the conditions and have calculated the runway length—you should be confident. Mental toughness does come into play. It is hard to just sit there and watch the obstructions at the far end getting closer and closer as the ground run continues. The temptation will be great to pull back on the wheel early in an attempt to muscle up and over those trees, but wait. If the elevator is deflected too soon, greater drag and a longer takeoff run will result. Keep the elevator neutral until the proper rotation speed is reached. After rotation, the nose of the airplane should be raised to a pitch attitude that will provide the best-angle-climb speed (Vx). Knowing what pitch attitude will accomplish this requires some practice. As a part of "getting to know" an airplane, you should spend some time maneuvering during slow flight. This is when we learn what pitch angle will produce what airspeed. This knowledge of the airplane must be put to work just after breaking ground during the short-field takeoff.

Once airborne, the pilot holds the Vx speed tightly. This speed will allow the airplane to climb the greatest number of feet in the shortest distance. After the danger from obstructions is past, the pilot should slightly lower the nose, accelerate from the critical speed, and climb at the best rate-of-climb speed (Vy). Best rate-of-climb will allow the airplane to climb the greatest number of feet over a unit of time. Remember these speeds; both Vx and Vy will not produce the maximum performance from the airplane if the airplane is not configured properly first.

Soft-Field Takeoff

I learned to fly on a hard-surface runway, so the first time I landed on grass I was surprised at just how bumpy it really was. Even the most well-kept grass strip will rattle and shake a small airplane on the take-off roll. It is easy to see why the soft-field technique was invented. Imagine a worst-case field where holes in the runway or rocks are hidden in tall grass. To take off in these conditions, a pilot would want to be rolling across the ground as little as possible and in the air as soon as possible. The longer the ground run, the greater the chances of damage to the airplane. The takeoff run over a soft or rough field begins using full-takeoff power and back pressure on the control wheel. Do not hold brakes, like when applying the short-field technique. Holding brakes or even slowing down anywhere might mean getting stuck. Roll right into the takeoff, and in a very short distance the elevator will be effective enough to bring the nose wheel off the ground. With the nose wheel in the air, there will be less drag from the wheels and less chance the nose wheel will hit something. An important part happens next. As the nose wheel rises into the air, the airplane continues to accelerate. This means that faster and faster air-flow is crossing the elevator, making it more effective. The pilot must now push the wheel forward ever so slightly so that the nose will not rise too high. If the pilot fails to do this, two bad things can happen:

1. The tail can strike the ground.
2. The excessive nose-high attitude can cause the airplane to make a premature lift off.

Remember the accident, reviewed earlier, where the passenger said that the airplane had become airborne after hitting a small ditch. Rough fields can have a tendency to "toss" you into the air before flying speed is reached. Once in the air, the pilot wants to make it fly. A soft-field premature liftoff is very hazardous. On hard surface runways, the same thing can happen when the airplane is loaded to the rear. The nose can come up too high and you are airborne before you are ready.

But the soft-field takeoff should take advantage of the ground effect. The soft-field takeoff is a tightrope. You cannot climb when the air-speed is too low, but you can fly—in ground effect. If the pilot can ride the cushion of ground effect while accelerating to a safe climb speed, then the wheels can be lifted off the ground early. When wheels/ground contact is lost, drag is reduced and the threat of striking something in the grass is eliminated. Once in the ground effect, pilots must be patient. If they attempt a climb here, the airplane will mush and most likely recontact the surface. The pilot again must have some confidence in what lift can do, given proper airspeed. The pilot after liftoff must lower the nose slightly to allow the airplane to accelerate through the ground effect. Then after reaching a speed where a climb from ground effect is possible, raise the nose to the Vx pitch angle. Most soft or rough runways are probably also short, so a Vx climb might be required following the soft-field takeoff technique.

Flap usage on soft or rough runways are often recommended by the airplane's manufacturer. Check the information handbook for the airplane you will use, because flap usage and settings are not universal. The use of flaps is a trade-off. Flaps will produce extra lift but also extra drag. If the main concern is getting bogged down or hitting objects during the takeoff run, then the extra lift that flaps can provide might be just what you need to get into the air fastest. But if clearing obstructions is the main concern, the extra drag that flaps produce might cut down the climb performance. It just depends on the desired result. The weight of the airplane might also have an effect on flap settings for takeoff. Consult the airplane's manuals and configure correctly.

Crosswind Takeoff

A crosswind adds yet another pilot challenge. Premature liftoff and loss of directional control have been the cause of several crosswind accidents. The crosswind takeoff is much like a normal takeoff, except that ailerons must be used in a way that prevents early liftoff. If the wind is coming from the right, the pilot should rotate the wheel to the right. This brings the right aileron up and spoils lift on the right side. Meanwhile the left aileron is down, which increases lift on the left. This holds the right wing down and prevents the wind from getting up under the wing. If the wind should get up under the wing, the right wing could lift off, the right main wheel will leave the ground, and the airplane will skip on the left wheel off the centerline. If the entire airplane gets into the air in this condition, the direction of flight will be toward the side of the runway, not down the runway, and the speed would be too slow to control the airplane.

The way to avoid the problem is to hold the upwind wing on the ground with aileron pressure. At the beginning of the takeoff roll, the airflow over the aileron will be minimal and therefore the deflection of the ailerons should be full. As the ground run continues and airflow increases, the aileron effectiveness will increase. The pilot should then gradually reduce the amount of aileron deflection, reaching neutral by liftoff speed. There are some extreme crosswind conditions where the aileron might not have reached neutral by liftoff, but ordinarily this should be avoided. If the ailerons are deflected as the airplane lifts off, then it will go into an immediate turn as it breaks ground. This could touch the wing tip to the ground just after liftoff.

The actual liftoff speed should be slightly faster than normal. Hold the airplane on the ground a little longer and then pop off the surface. Coming off a little faster helps in two ways. A slow liftoff in a crosswind can blow the airplane downwind and toward the runway edge like a leaf. And, with more speed, the airplane will jump into the air, yielding enough immediate altitude to begin a crab angle. If you turned into the wind just over the surface, the upwind wheel could touch down again as the wings were initially banked. Getting higher sooner allows the crab angle to be established, and the airplane can climb out right above the runway centerline despite the crosswind.

Wake Turbulence Avoidance

One of the by-products of lift is wake turbulence. As illustrated in Figure 5.4, the wing's airfoils create a high-pressure zone on the underside of the wing and a low-pressure zone on the top. This differential pressure produces the force that makes airplanes fly and that we call lift. In nature a high pressure always wants to equalize by flowing to a low-pressure area. This is why when you blow up a balloon and then let it go, the high-pressure air inside comes rushing out and the balloon flies around until the pressure is the same inside and outside. The high-pressure air under the wing wants to equalize and fill in the low pressure, so the high-pressure air pushes up. But out on the wing tip, the high pressure finds an easier way to the low-pressure area by sneaking around the corner. The air curls around the wing tip. This takes place while the airplane is in forward motion, so when the air curls around it is also left behind. The result is a corkscrew or vortex of air that trails behind the wing tips. If you could see it, it would look like a horizontal tornado. The greater lift that is produced, the greater the effect.

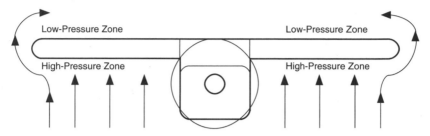

Fig. 5.4 The production of wing-tip vortex.

Every airplane, even small radio-controlled airplanes, produce wing-tip vortex. But when the airplane is large, the effect can be dangerous to smaller airplanes.

The greatest danger from wake turbulence/wingtip vortex is from large airplanes in slow flight. In slow flight, large airplanes must produce extra lift to make up for the slow speed by using flaps, leading-edge droops, and a range of other devices. Another airplane caught in this vortex can be thrown out of control. When the vortex is generated, it will tend to sink, and spread out. If there is a crosswind, the vortex will stay in tact and move in the direction of the wind.

When flying behind another airplane, stay well back and if possible slightly above the other airplane's altitude. Wait at least three minutes when taking off or landing on the same runway that a larger airplane has just used.

Killing Zone Survivor Story

NASA Number 418103

After completing my runup and checklist, I waited for another aircraft to depart. I pulled onto the runway and departed <u>immediately</u> after the aircraft in front of me. The wind was from my left, at approximately 10 knots. After departure, I felt the aircraft sinking. I immediately dropped the nose and turned away from the flight path of the aircraft I was following. As my airspeed increased, I pulled the nose of the airplane up and turned back into the wind. I began a positive rate of climb. During the pull-up I noticed that I came within approximately 50 feet of the airport's rotating beacon. Onlookers commented on me buzzing the office, but that was not the case. Apparently, my takeoff was too soon after the aircraft in front of me, and the crosswind and turbulence caused an abrupt loss of lift. As a result of this experience, I am a firm believer in wake turbulence, even behind a small aircraft. I was in no hurry to get anywhere, and should have waited a little longer before takeoff. This boils down to a very poor decision on my part.

Use all these takeoff techniques together with your airplane's information manuals and you will make safer takeoffs. Every takeoff is different, even if you always fly from the same airport. Each takeoff will have different wind, different density altitude, and different airplane loading. Plan each and every takeoff as if it were the first in that airplane. Doing so could have prevented most of the takeoff accidents in the category.

CHAPTER SIX

Approach and Landing

W HEN TAKEN TOGETHER, takeoffs and landings do not make up a large percentage of the total flight time, but combined they are the most accident prone of any other groups. Between the two—takeoffs and landings—the landings are the most hazardous. Figure 6.1 again illustrates a clear Killing Zone of fatal landing accidents beginning at 50 hours and ending at approximately 350 flight hours. These numbers include accidents on full-stop landings, stop-and-go landings, touch-and-go landings, and go-arounds or rejected landings. There were 1875 fatal landing accidents in this group from 1983 to 2000. Why would landings be more dangerous than takeoffs? There could be several reasons. Takeoffs start slowly and accelerate because of engine power. Landings already have speed and can accelerate because of gravity. Takeoffs are normally traveling away from the ground. Landings are normally traveling toward the ground. One major difference is that takeoffs occur at the beginning of the flight when the pilot is more rested. Landings occur at the end of the flight when fatigue, boredom, and complacency have set in. The physical act of landing may not be more hazardous than the act of takeoff, but it takes place when the pilot is the least rested and possibly the least alert.

NTSB Identification: ATL94LA174. Linden, Tennessee

The Private Pilot was conducting a touch-and-go landing. After touchdown, the flaps were retracted and power was added for the go-around. The engine hesitated, then provided normal

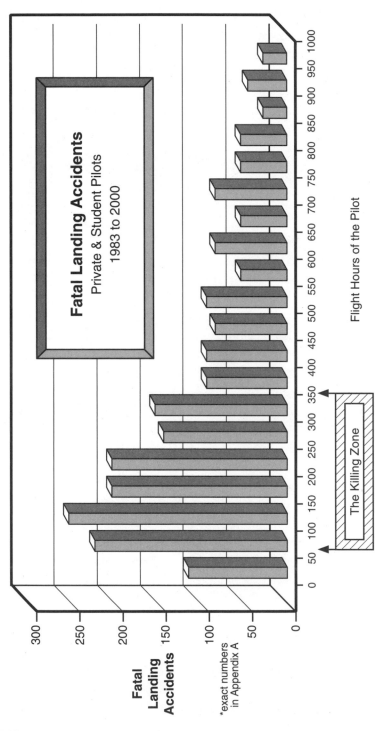

Fatal Landing Accidents
Private & Student Pilots
1983 to 2000

Fatal
Landing
Accidents

*exact numbers
in Appendix A

The Killing Zone

Flight Hours of the Pilot

Fig. 6.1

power. As the speed of the airplane increased, it began veering left. Right rudder was applied without result. The pilot then cut power as the airplane overran the left side of the runway, continued down an embankment, struck trees, and nosed over. Post accident examination of the airplane revealed no anomalies with the engine and continuity of the flight controls. There was evidence of side loading on the outboard side of the right main tire.

Probable Cause:

The pilot's improper use of the rudder that resulted in a loss of directional control.

Figures 6.2a through 6.2d are all from this accident. The pilot was unhurt, but the airplane was substantially damaged. 6.2a shows the airplane tail upside down and sticking up into the air. The runway is just beyond the horizon in the photo. When the airplane "veered" left and overran the left side of the runway, it came over that hill and down into the ditch. 6.2b also shows the airplane as it came to rest. The runway is off to the right of the photograph. The airplane came from the right, down the embankment, and finally the nose wheel dropped into the ravine and the momentum took the tail on over. It's amazing how much trouble you can get into when you don't do things right.

Look carefully at Figure 6.2c. This photograph was taken back up on the runway. You can see the skid marks left by the right main wheel. The report said, "there was evidence of side loading on the outboard side of the right main tire." The skid marks show the angle at which the airplane departed the left side of the runway. The embankment that the airplane ultimately rolled down is beyond the grass and short of the far trees. Notice that the skid marks cross the taxiway exit lines.

Figure 6.2d is an aerial photograph of the crash scene. The pilot was making the touch-and-go landing from left to right in the photograph. You can see the airplane (center) off to the left side of the runway. You can also see the taxiway exit lines. Picture the pilot's path from the runway just beyond the taxiway, across the field, and down the hill.

The NTSB report also said that, "After touchdown, the flaps were retracted..." but you can see in Figures 6.2a and b that the flaps are still partially down. The pilot had said that power was cut before the airplane went down the hill, and when the nose wheel dropped and the tail came over that, the engine had stopped. The mixture control was found

Fig. 6.2a After departing the left side of the runway, the airplane rolled down a hill and into the ravine.

Fig. 6.2b The airplane rolled down the hill from right to left in this photo.

Fig. 6.2c The bold white strip that crosses left to right in the photo is the edge of the runway. The right main tire's skid marks show where the airplane went out of control just past a taxiway intersection.

in the aft, idle cutoff position, and the damage to the propeller was consistent with a stationary impact.

What really happened here? Why did the engine hesitate and then regain normal power? Did the power hesitation distract the pilot and contribute to the loss of control? A power hesitation from a carbureted engine is normal on touch-and-go landings as well as stall recoveries and go-arounds when the throttle is pushed forward too quickly. Figure 6.3a and b illustrate what happened. Figure 6.3a depicts the typical airplane updraft carburetor when on final approach to land. The throttle valve is all but covering the passageway to the engine. With the path

Fig. 6.2d The airplane had been landing from left to right. The taxiway intersection where the airplane left the runway and the airplane itself can be seen in this aerial view of the accident.

mostly blocked, less air is drawn into the engine. The venturi is the narrow portion of the diagram, and it is here that a low pressure is produced. The air must flow faster to get through the narrowing passage. When the air accelerates, the pressure drops and this draws fuel into the venturi. The fuel is drawn, just like drinking from a straw, into the air through the discharge nozzle. The nozzle has a fine screen that turns the liquid fuel into a mist. The air and fuel mix together here, and then travel to burn together in the engine's cylinders. The amount of fuel and air is "metered" so that the proper ratio of air and fuel get together. The ratio varies but it usually is around 12:1. That is 12 parts of air for every 1 part of fuel. If the engine burns 10 gallons of fuel in an hour, then approximately 120 "gallons" of air is also burned with the fuel. The ratio is very important. If there is not enough fuel in the ratio, say 18:1, then the mixture is said to be "lean." There comes a point where the mix can get so lean that the engine stops altogether.

Figure 6.3b is the same carburetor, but now the throttle has been pushed in to the "wide open" position. You can see that the throttle

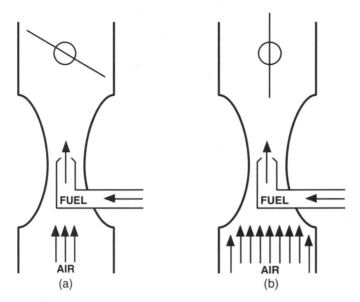

Fig. 6.3 (a) The carburetor with the throttle partially closed at idle. (b) The transition to throttle wide open.

valve has turned and no longer blocks the way to the engine. Air comes rushing in. As the airflow increases, the low pressure in the venturi gets even lower, and this in turn will draw more fuel in. In theory, the ratio should stay the same even as the air increases. When the amount of air goes up, the venturi action should draw additional fuel in, and the mixture ratio is maintained. But in practice, if the throttle is opened too quickly the air rushes in before the fuel can catch up. This momentarily leaves the engine with an extremely lean mixture. The mixture can get so lean that the engine will "hesitate" or "sputter" or just quit.

The pilot involved in this accident became anxious during the touch-and-go. He had landed long and fast. In the back of his mind he knew the runway end was quickly approaching. In his haste to get back in the air before running out of runway, he jammed the throttle forward. In that moment the throttle valve went wide open, the air came rushing in at a faster rate, but the fuel was still coming in at the slower rate. The engine hesitated. The pilot froze. In the next moment, the additional air flow created an additional low pressure in the venturi, which drew additional fuel into the mix. This restored the proper ratio. It took another moment for the air/fuel mixture in the proper ratio to reach the engine. When it did, the engine "caught" and the engine RPM accelerated to full power in the blink of an eye. The engine torque went from near zero when the engine hesitated, to near maximum torque as the engine went to full power. The sudden and unexpected increase in torque veered the airplane to the left. The pilot did not react quickly enough, and soon he was off the runway and headed for the ravine.

The solution: Always move the throttle smoothly. Let the air and fuel amounts rise together. This will ensure even and continuous power. Moving the throttle too quickly could even stop the engine altogether when you need power the most, like on a rejected landing. It pays to know how things work and to remain cool under pressure. This accident could have easily been avoided. Fortunately, this pilot was not injured, but others were not so fortunate.

NTSB Number: LAX86FA282. Phoenix, Arizona

The pilot performed one normal touch-and-go operation, and touched down normally. [The accident occurred] on the second planned touch-and-go which was to terminate with a right departure out of the pattern. After touchdown, witnesses observed the aircraft roll straight along the runway centerline for about 800 feet then it began a gradual divergence from the runway which seemed to stabilize at about a 30 degree angle. After departing the runway edge, the aircraft rolled at a constant speed for an additional 900 feet through the dirt area between the runway and adjacent taxiway before colliding with a concrete drainage abutment. The witness saw the wings of the aircraft rock back and forth during the ground roll through the dirt. Examination of the aircraft tracks confirmed that the aircraft did oscillate about the roll axis and shift weight from side to side during the ground roll. The pilot had accrued only eight dual hours to the first unsupervised solo. The accident flight was the second solo flight without supervision. No preimpact failures of malfunctions were found.

Probable Cause

The pilot's improper use of the rudder that resulted in a loss of directional control.

The two accidents sound much alike, but the second, tragically, killed the student pilot. Complicating factors was the mention of this student's "unsupervised" solo flight. This means that the flight instructor who had originally signed off the student to solo was nowhere to be found. The report also makes note that the student had only eight hours of instruction prior to the first time he soloed. The regulation (61.87) does not require a minimum number of instruction hours before a solo flight, but does require that the student become "proficient" in all areas from a list of 15 categories with a total of over 40 items. It is doubtful that the student could have truly been proficient in over 40 tasks within just eight hours. The report is implying that the flight instructor was at least lax or at worst negligent in instruction and supervision of the student. Flight instructors should make it very clear and student pilots should understand that when they are signed off to fly solo, that endorsement can only be used with the permission and consent of their own instructor. It is not known if the instructor in this case made that clear to the student, or if the student was just flying against the known wishes of the instructor.

I had a student once who, with his private pilot brother, owned their own airplane. One day when I arrived at the airport, I noticed that their airplane was gone from its normal parking spot. I assumed that the brother was out flying, but it started to bother me; could it be that the student pilot, not the private pilot, was out flying without my knowing about it? The two brothers were in business together, so I decided to call their office and ask to speak to the student pilot. The phone range once and the brother/private pilot answered. My worst fear had been confirmed. The student pilot—my student—was in fact flying around using my endorsement! I was so nervous that I actually went to the empty parking space and paced until he returned. When he arrived, all smiles, I asked him what he thought he was doing. "Had to run over to xxxxville to get some paperwork," he said. At this point I am listing off in my head all the violations that had been committed:

1. He flew across country with no cross-country endorsement.
2. He flew without my knowledge and consent.
3. Student pilots cannot fly in the furtherance of a business and it sounded like getting "some paperwork" was a business deal to me.

I quickly realized, however, that most of the blame was my own. I had never actually come out and said in so many words, "You cannot fly, even if its your own airplane, unless I say it's okay." To my knowledge he never flew solo again without it being part of his training and with me present and supervising. But I always looked over at that parking space when I first got to work until he got his private certificate. If my student had an accident during his unauthorized solo flight, I would have been in the same situation as the instructor of the student in the previous accident example. Just like my situation, it could have been a big miscommunication, but in this case it was a fatal miscommunication.

So many skills are required for safe landings. Often these skills must be performed simultaneously. The pilot must have good vision, must have good visibility, and must use good judgment when it comes to terrain, turbulence, and winds. The next set of accident examples prove that consideration of all of these factors is required for a safe outcome.

The Lost Eyeglasses Landing

NTSB Number: LAX89DUG03. Hesperia, California

Witnesses reported observing the pilot taxi to the runway with the canopy cracked open approximately two inches. The pilot entered the runway and departed without pretakeoff or cockpit check. Soon after takeoff the canopy departed the aircraft along with the pilot's glasses that were both found on the runway. The pilot returned to land and the aircraft entered into a stall and spun into the ground from an altitude of 100 feet about one quarter mile from the runway.

Probable Cause:

The pilot's failure to properly latch the canopy prior to takeoff and by allowing the aircraft to enter into a stall during a critical phase of flight at an altitude insufficient to recover safely. A factor in the accident was the loss of his eyeglasses when the canopy departed the aircraft.

This private pilot held a valid third-class medical, but it had eyesight restrictions, requiring the pilot to wear corrective lenses.

The Sun Glare Landing

NTSB Number: LAX 91FA012. Goleta, California

The pilot was instructed [by ATC] to enter a left base in the traffic pattern and land on runway 25, for the purpose of maintaining separation from other landing traffic. The air traffic controller additionally instructed the pilot to make the base leg square. A review of the airplane's flight path as depicted by FAA drawings, illustrated an approach pattern similar to a long final approach. Witnesses on the ground about one and one half miles from the airport on runway 25 extended centerline stated that the airplane overflew their homes at a low altitude. There were no reports of unusual engine noises. Air traffic controllers observed the airplane in a rapid descent immediately before the accident. There was no evidence of mechanical failure or malfunction with the airplane before the accident. The azimuth and angle above the horizon of the sun was 241.1 degrees and a 2.8 degree respectively.

Probable Cause:

The pilot's failure to maintain proper glide path while on final approach. A contributing factor was sunglare.

The runway number was 25, which means that it aligned within 10 degrees of the 250-degree heading. The sun was reported at 241.1 degrees. This means that the sun was directly or near directly in the pilot's eyes on final approach. In addition the sun was low, only 2.8 degrees above the horizon. This proved to be a critical lesson for us all: You cannot land on something that you cannot see. There are many reasons why you should reject a landing and go-around. In this case it was direct sunlight. Ironically, if the sun was 2.8 degrees above the horizon, at the rate the earth turns, the sun would have set in approximately 10 minutes—roughly the amount of time to execute a go-around and make a second attempt at landing.

The Crosswind Landing

NTSB Number: CHI96FA117. Carrollton, Ohio

Witnesses observed the airplane in an attempt to land on runway 7. The wind was reported to be from 180 degrees at 15 knots with gusts to 20 knots. Witnesses describe the airplane initiating a go-around and immediately turn to the north in a near vertical bank. Witnesses then reported the airplane descended through the roof of a hangar and exited through the rear of the building. Subsequent examination of the airframe and engine failed to reveal any preimpact anomalies. The landing flaps were found in the 30-degree extended position.

Probable Cause:

The pilot's inadequate compensation for the crosswind conditions, his failure to maintain directional control, and his failure not to raise the landing flaps during the go-around. The crosswind was a factor.

Landing on runway 7 with a wind from 180 has both a crosswind and tailwind component. The airplane was a Piper Cherokee 180. The pilot had 58 hours in that airplane and 80 hours total. This private pilot and one passenger were killed. Runway 25 would also have been a crosswind, but it would have been a more conventional headwind/crosswind component set up. Most pilots don't think about an alternate airport on VFR days. But the weather can be great and the wind be terrible for the runways at your destination airport. Get familiar with the runway alignment of airports that surround either your home airport or your destination airports. There will come a day when the visibility is very far and the clouds very high, but you still divert to another airport, just to get a better wind angle on landing.

The Tailwind Landing

NTSB Number: FTW92FA098. Oklahoma City, Oklahoma

The pilot landed downwind on a 3,350 foot runway. After touching down beyond the midfield point, he elected to abort the landing. A nose high attitude was established on climbout and the airplane stalled as the pilot initiated a left crosswind turn.

Probable Cause:

The inadvertent stall during aborted landing. Factors were the tailwind, the pilot's selection of the wrong runway for existing conditions, and his long landing.

Today most uncontrolled airports have an AWOS weather reporting system that continuously broadcasts the wind direction and velocity. Use this service every time it is available. If it is not available, use the "old-fashioned" methods: wind sock, wind tee, or tetrahedron.

The Turbulence Landing

NTSB Number: CHI97FAI 16. Holland, Michigan

According to one of the passengers, during the flare/touchdown as the airplane got down near the runway, there was considerable turbulence, "the airplane traveled down the runway without landing for a considerable distance, when the [pilot] decided to go around." He said that as they passed the end of the runway, he felt the tail of the airplane was low and, "...they seemed to be floating just over the tops of the trees with wiggling type motion by the plane." He said the pilot said that, "the airplane won't climb," and they were barely clearing the trees. The left wing went down, the airplane hit the trees and impacted into a garage roof. Prior to the flight the pilot had told one passenger that he had never flown the airplane with a passenger in the rear seat that was the case on this flight. A post-accident examination of the airplane and engine revealed no preexisting anomalies.

Probable Cause:

The pilot's failure to maintain adequate airspeed that resulted in a stall/mush and subsequent impact with trees and then a residence. A related factor was turbulence.

Sometimes you will hear pilots say, "keep a little extra airspeed on final when it's turbulent," and this is good advice in general. The idea is to maintain lift during the approach with a little extra speed so that when the turbulence hits, the airplane will not be as close to stall speed. But this can be carried too far. If the airplane has excessive speed, the airplane will not land on schedule but float down the runway instead. This accident was in a low-wing airplane, which would have added to the ground effect float, but it can happen in any airplane. While floating down the runway, and as the airspeed begins to dissipate, the airplane will be even more vulnerable to wind gusts and turbulence. And as the airplane coasts to ever slower speeds, the safe go-around possibilities diminish. A good rule of thumb for turbulent approaches is to add approximately 5 knots to normal landing speed but no more.

Be ready for turbulence close to the ground on short final. If there is any wind, the air will be more choppy near the surface, where the airflow interacts with objects on the ground. Be especially wary of a crosswind flowing through a row of trees that line the edge of a runway. The air will come spilling out of those trees and jolt the airplane at about flare altitude.

The Hard Landing

NTSB Number: LAX92FA196. Corona, California

A student pilot was in the left seat of his newly purchased aircraft [Beech A23-24] and a private pilot occupied the right seat. The investigation was unable to determine who was flying the aircraft at the time of the accident. The pilots initiated a go-around after a hard landing that sheared off the left main landing gear, broke the nose gear strut, and damaged the horizontal stabilizer. Witnesses stated that the climb rate and airspeed appeared to be very low, and that the left horizontal stabilizer appeared to be bent down. According to witnesses, during the crosswind and downwind legs, the airplane was making abrupt pitch changes and never attained altitudes of more than 300 to 500 feet AGL. While on a close downwind leg, the airplane struck an industrial building that is approximately 100 feet higher than the runway surface.

Probable Cause:

(1) The flying pilot's improper landing flare that resulted in a hard landing and significant damage to the aircraft empennage flight controls, (2) the flying pilot's decision to initiate a go around with obvious serious damage to the aircraft, and (3) the flying pilot's inability to control the aircraft in pitch due to the damage sustained in the hard landing. A factor in the accident was the pilot's lack of experience in the accident airplane.

The report refers to the "pilot flying" but it was not clear which pilot that was. If the student was flying the airplane, then he was illegally carrying a passenger. If the private pilot was flying the airplane, then he was flying an airplane that he was completely unfamiliar with and doing so from the right seat. Remember that takeoffs are optional. With hindsight it would have been better to semicrash land the damaged airplane rather than attempt to make it fly again. Any unusual landing demands an inspection before flight again, even if the inspection is done by the pilot. The decision to go around here happened in a fraction of a second and in the midst of confusion. A landing hard enough to cause that much damage had to have been loud and jarring to the pilots. More confusion may have resulted when two pilots tried to make the same decision at the same time. It is possible that one pilot made the hard landing and the other pilot attempted the go-around. There can only be one true pilot in command. In this case, flight in an unfamiliar airplane, with inexperienced pilots, led to a faulty landing and a deadly, split-second decision.

The Unfamiliar Airport Landing

NTSB Number: CHI91FA216. Seneca, Illinois

The pilot was on short final for landing at a private airport and struck an unmarked powerline. The aircraft crashed into a creek short of the runway. Witnesses described the approach as low and flat. The powerline was 28 feet above ground level, and 485 feet from the runway threshold. Examination of the pilot's logbook revealed no previous recorded flights to this airport.

Probable Cause:

Proper glide path not maintained, inadequate altitude, and inadequate visual lookout by the pilot in command.

The pilot had 140 total hours of flight time. The pilot was killed. Two passengers survived with serious injuries. When planning a flight to an unfamiliar airport, you should gather any and all information you can. Many airport directories are available and these come with notes attached. These notes include advisories about intense student training, or deer on the runway, or trees and power lines. This accident happened at a private airport. Private airports often are not included in such directories, so in those cases call ahead and speak to the runway owner about the best approach, runway slope, surrounding terrain, and power lines.

The Aborted Landing—Off Runway—Go Around

NTSB Number: LAX88FA306. Temecula, California

After landing his aircraft on the 3,023-foot asphalt runway, the pilot lost directional control and the aircraft ran off the left side of the runway approximately 635 feet from the threshold. The pilot was able to regain control of the aircraft and initiated an aborted landing from the off-runway location. The aircraft overflew another aircraft that was approximately 200 feet ahead of the point where the aborted landing was initiated. Then the aircraft collided with a metal ladder attached to a 24-foot high commercial building in his flight path.

Probable Cause:

Failure of the pilot to properly initiate a go-around during an aborted landing. The loss of directional control during the landing, failure to return to the runway for the takeoff, not establishing the proper rate of climb due to density altitude, and failure to raise the flaps are considered factors in the accident.

This accident had more poor decisions per second than any other cases used here. The pilot was first landing too close behind another airplane on a runway that was barely over 3000 feet long. The pilot lost control of the airplane after landing and ended up in the grass. This was bad, but the airplane was undamaged and the pilot was unhurt. But for some unexplained reason the pilot felt that he needed to make a touch-and-go from the grass along the side of the runway. The takeoff was made without retracting the flaps from the previous landing. Then this pilot endangered others by overflying the other airplane and striking an object on a building.

Many of the accident cases in this chapter involve pilots in the process of making a touch-and-go landing. I teach at an airport that has 3900 feet of runway and I don't let my student pilots make touch-and-goes anymore. I strongly advise other pilots not to do it either. This maneuver just has too many accidents associated with it for me to be comfortable. Too many things have to happen just right, during a very short amount of time, for it to be a safe maneuver. I once talked to a very nervous pilot flying the traffic pattern after a touch-and-go. The problem was that when he raised the flap position handle during the ground roll, the flaps did not come up. He rotated without noticing that the flaps were still full down. Needless to say the airplane did not climb as it was supposed to and he made a tree-top level downwind before making a safe landing. The flaps on his airplane were operated by an electric motor and the circuit breaker for that motor had popped out. The circuit breaker may have been out for the duration of the flight and he did not inspect it before landing (the checklist calls for an inspection of all circuit breakers), or it may have popped when the flap handle was adjusted. Either way, if he had been making a full-stop landing, the inoperative flaps would have been detected before the next takeoff. Instead he was back in the air in a suddenly unairworthy airplane. He managed to land safely, but we have seen examples of others that did not.

Landing Tricks and Traps

Normal Landing

Part of getting familiar with an airplane is learning what combinations of pitch, power, and flap settings come together to produce desired airspeeds. The normal landing is a systematic configuration change from the cruise phase of flight to touchdown and rollout. Each airplane is somewhat unique and we saw from several of the previous accident examples that inexperience in a particular airplane can be an accident cause. So do not fly an airplane until you are really comfortable with its power, cockpit arrangement, and speeds. Insurance companies require a "check-out" flight for pilots before they will ensure the pilot. The length of the check-out and what is exactly covered will vary among insurance and rental companies. The check-out may not be enough to be completely prepared. When you check out, ask many questions, get into the airplane manuals, and fly more than the check-out requires if that's what it takes to be really ready.

The normal landing is an art form. You must allow the airplane's speed to dissipate in the flare so that sufficient lift is no longer being produced at the moment that the wheels reach the ground. If the timing of this is off, then the landing will be more exciting than necessary. If you arrive at the surface with too much speed, the airplane will not be ready to land and you will float. Floating makes the pilot vulnerable to many problems, like crosswind, turbulence, and the end of the runway. If you reduce the speed before you reach the surface, then you could experience the opposite of float: the drop-in. If the airplane stops flying when you are still 10 feet in the air, the airplane will fall the rest of the way. This can produce a rattling, possibly damaging hard landing. So the two variables, speed and altitude above the surface, must come precisely together.

The key to good, smooth touchdowns are consistent, stabilized approaches. A good landing does not just happen by accident; it happens because of good preparation. If you can prepare the airplane each time from the traffic pattern and into the final approach with a set approach angle, a set amount of flaps, and a set airspeed, then practice will make the touchdowns smooth. But if you come in each time at a different angle, which requires ever changing flaps while wrestling with erratic airspeed, the touchdown will be an unpredictable adventure. Getting to the point of having a stabilized approach each time will not only help the normal landings, but also will help make the variations necessary to meet the demand of "specialty" landings like those on soft fields and short fields.

The Short-Field Landing and LAHSO

The short-field landing is a required maneuver for every airplane checkride from recreational through ATP. The majority of flight training takes place from runways that are in fact long enough for the type of airplanes involved, so our short-field landing practice has become more a simulation than a practicality. When maneuvers become "make believe" in our minds, we soon don't take them seriously, but short-field landings are not a simulation. You may not ever venture to a truly short runway and face an actual short field, but every pilot faces LAHSO. This is short for Land and Hold Short Operations. LAHSO takes place at larger airports that have intersecting runways. The air traffic control managers came up with LAHSO in a stop-gap attempt to provide more available runways to our overcrowded system. When land and hold short operations are under way, controllers will give permission to land to two airplanes flying to different runways. The problem is that those runways cross. That would mean that if both airplanes rolled out after landing as normal, they could collide in the intersection. The solution: One airplane must stop prior to crossing the intersection. One airplane must land and then hold short. This is where our short-field landing practice must pay off.

The runway you are cleared to land on may be 6000 feet long, but you may be authorized to use only the first 2000 feet. You cannot land long. You cannot come in too fast and float. You must get down and stopped with little room for error. To handle LAHSO safely, you must first know how much runway you have been given. The controller will only say, "Cleared to land runway 32, hold short of 19." LAHSO available runway lengths are published in the back of the FAA's *Airport/ Facilities Directory* for every controlled airport with intersecting runways. If you plan to fly to a controlled field that has crossing runways, you should look this information up before departure. But if you are given a LAHSO instruction and have not checked the distance, then ask the controller to tell you the distance. When the available runway length is known, you must compare that with the distance that your airplane requires. If there is sufficient distance, then put your short-field technique to work. If there is not enough space, tell the controller that you will need to come back when you can have the full length. Be advised, though, at a very busy airport, the next time the full length might be available could be midnight.

Unfortunately, general aviation pilots do not have a good reputation when it comes to dependable short-field landing techniques. The FAA completed a study some years ago (DOT/FAA/CT-83/34) that tested private pilots for skills retention. The FAA followed a group of new private pilots and gave them evaluation flights at 8-, 16-, and 24-month intervals after they first passed the private checkride. On the day of their checkride, 90% of the pilots met FAA standards when performing short-field landings. But by the time 24 months had come and gone, only 51% were able to meet private pilot standards. That is correct. More than half the private pilots could no longer handle the short-field landing properly and/or safely. That was information that the Airline Pilot Association (ALPA) was aware of even without the FAA study. The ALPA members make up the majority of the U.S. airline pilots. They voted to reject all takeoff and landing clearances when general aviation airplanes are given a LAHSO instruction. This means that the airline pilots simply do not trust the skills of the general aviation pilots. They will not approach an airport on a converging course with a GA pilot because they fear that the GA pilot will mess up the short-field landing, roll out across their runway, and a terrible accident will result. The controllers now know not to give LAHSO when it is an airline/GA combination. This means that quite often the GA airplane will be delayed, turned out, given a vector, to allow the airliner to get down and off the runway. We all get hurt because as a group our short-field techniques are suspect.

The short-field technique requires critical speed, flap, and power control. Commonly, full flaps are used to provide the steepest approach. In practice, we simulate a row of trees or a power line at 50-feet height. The object is to clear the obstructions and yet be able to touch down without rolling off the far end of the runway. The perfect technique would be a short final with the airplane's landing gear down, full flaps, and proper short-field airspeed. Reduce the power gradually, eventually to power off. The airplane just clears the 50-foot obstruction and immediately thereafter begins a flare and normal touchdown. Once on the surface, the weight is quickly transferred from the wings to the wheels by retracting the flaps. This makes the brakes more effective and the airplane comes to a stop without skidding. That is the perfect technique, but many common errors take place.

The approach should begin from the base leg on higher than normal. You must be deliberately high on final approach so that you can use full flaps and therefore get the steep approach. All too often, pilots make a normal approach, and then when full flaps are applied the airplane gets too low. The pilot then responds to being low with power. Now the pilot has full flaps and high power, struggling to make to the runway. It's like applying brakes (flaps) and gas (power) at the same time in your car. It doesn't make sense to have to step on the brakes and the gas simultaneously, but this is what happens when you don't fly the short-field approach high to begin with. If you do arrive at the runway with a very slow speed and a very high power setting, the airplane will literally fall out from under you when the power is reduced. This probably will make a short landing, but it's hard on the airplane, hard on the occupants, and is not a safe technique. Stay high and then use full flaps because you need full flaps to make a steep approach.

Pilots then trade off. They are coming in over the obstruction just fine, but they notice the airspeed is too slow. They lower the nose to gain back the airspeed, but now they will be too low to clear the obstruction. In this case the nose should be lowered to yield the proper airspeed, but also a small amount of power should be added to adjust the approach angle. A small amount means approximately 100 to 200 RPM or 2" to 3" of manifold pressure.

Another problem is the touchdown. Often the pilots will come in slightly fast and they see their target touchdown spot coming up quickly. In an effort to "nail it," they push forward on the yoke and force the airplane to the ground on the spot desired. The problem is that when you push forward on the wheel you leave the landing configuration: The nose wheel moves closer to the surface, with the possibility that the nose wheel might touch the surface first. When pilots come in too fast, they try to erase their mistakes by "planting" the airplane down before the airplane is ready to land. Again, this is hard on the airplane, the occupants, and can be more dangerous than the short landing in the first place. If the speed has been controlled properly before the flare, the airplane will dissipate its speed naturally and land on the desired spot with a normal touchdown.

Overbraking is also a common problem. If too much brake pressure is applied, the wheel will "lock up." Our airplanes do not have antilock brakes like many cars do these days. When the brakes are held so tightly that the wheels just stop turning, they will skid, shudder, and bounce. This actually increases the stopping distance. During the moments when the tire is bouncing off the ground, there is no braking action at all. When the tire is on the surface it will scrape across and the heat will melt the rubber. This causes a slick or bald spot on the tire tread. Brakes should be applied evenly with all the pressure to stop without skidding. The brakes themselves can overheat during and after heavy braking. I once landed and, knowing that a faster airplane was just behind me, I hit the brakes hard to slow down in time to make the next taxiway. After getting off the runway, I happened to look down at the left brake. It was nighttime and because it was dark I could see the orange-hot glow of the brakes. They were much hotter than I ever expected. I also know that I had overbraked before in the daytime but did not notice how hot the brake had gotten. The metal brakes can get so hot that they can burn. And a metal brake fire is a real emergency. What is usually situated just above he wheels in the wings? The fuel tank! If you overdo it and the brake starts to burn, the tire will blow out, and the little fire extinguisher that might be in the airplane will never put out a metal fire. You will have to evacuate the airplane, and the entire airplane will burn. I saw the remains of a new airplane that had been burned up after a hard brake lit off the brakes. It is a shame to lose an airplane because of poor short-field landing technique.

How can you tell how much runway is enough runway for a short-field technique? The manufacturer will provide a table or chart that can be used as a guide. Each airplane and each manufacturer will be different, but the runway markings are standard. Figure 6.4 depicts the standard distances for a nonprecision approach runway. On the right side of the figure are the lengths of each marking and at the far right is the cumulative distance from the runway threshold. The distances are standard. The digits of the runway numbers are 60 feet long. The runway centerline strips are each 120 feet long with 80 feet between the stripes. Somewhere near the middle of the runway you could find at least one stripe that was not 120-feet long to accommodate runways of various lengths.

On the left side of Figure 6.4 is a sample short-field landing chart. Your airplane's charts may not look like this, but the same information will be presented on either a table or chart. This example shows that the total distance to land over a 50-foot obstruction is 1375 feet. This includes flying over the obstruction, flaring and touching down beyond the obstruction, and braking to a complete stop—1375 feet in total. The table also gives a distance from "ground roll." This means that of the 1375-foot total, 655 feet of that will be spent on the ground braking to a stop. With this information you can calculate how much of the total will be spent in the flare before touchdown. Total distance (1375) minus the ground roll (655) equals the distance of flight from over the obstruction to touchdown (720).

In order to properly determine my performance, I simulate that the obstruction is sitting exactly on the runway threshold. This means that I would cross the runway threshold at least 50 feet in the air. I do not shoot to land "on the numbers" because it would be hard to determine how far back from the numbers to place the obstruction. I use the runway itself as my yardstick to determine the proper touchdown spot. Referring to the standard runway distances of Figure 6.4, my touch-

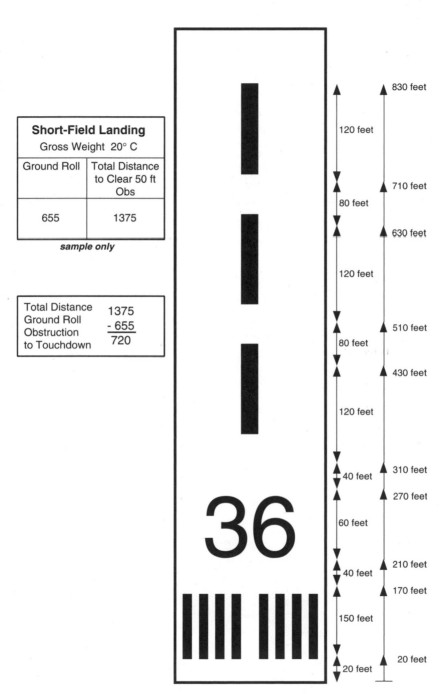

Fig. 6.4 Standard runway markings and their distances.

down spot should be just at the beginning of the third centerline strip. The strip itself begins 710 feet from the obstruction (threshold), so my calculations place the touchdown in that third stripe. The ground roll begins from the touchdown until wheels stop. The total distance, or the wheels-stopped position of 1375 feet is beyond the distance shown in Figure 6.4. Manufacturers will include with landing performance charts provisions for landings against a headwind or on rough surfaces as well.

The commercial pilot checkride standards that are published in the Practical Test Standards (PTS) describe what the FAA thinks is adequate short-field technique. With regards to the touchdown location the PTS says, *Touchdown at a specified point at or within 100 feet beyond the specified point, with little or no float, no drift, and with the longitudinal axis aligned with and over the center of the landing surface.*" The PTS refers to a "specified" point. This specified point is the spot on the runway that the pilot, in this case checkride applicant, says will be the touchdown spot. You can use the calculation method from Figure 6.4 to determine the "specified" point. The PTS gives the tolerance for maneuver at $-0/+100$ feet: "*...at or within 100 feet beyond the specified point.*" Using the beginning of the third centerline stripe as the example, a touchdown anywhere in that stripe passes the checkride.

The Private Pilot Practical Test Standard for short-field landing states: "[the applicant must] *Touch down smoothly at the approximate stalling speed, at or within 200 feet beyond a specified point...*" So private pilots must not land short of the target, but have twice the distance window required of commercial pilots.

Passing the checkride is important, but safely flying an actual short landing or LAHSO is even more important. Look over the landing tables and charts of the airplanes you fly. Become familiar with the landing distances that can be expected from the airplane and go out and practice. This way the next time a controller says, "...cleared to land runway 9, hold short of 36," you can accept or reject that clearance with confidence.

Soft-Field Landing

The soft-field landing technique assumes that you are truly landing on a rough, soggy, or soft surface. The idea of the soft-field takeoff (Chapter 5) was to get into the air as soon as possible, so that if there are any rocks, ruts, or holes in the runway the airplane will not strike them. The same idea holds for a soft-field landing. We assume that we could hit something hidden in the grass or dirt that could damage us or the airplane. If we do hit something, the damage would be reduced the slower that we are traveling when the impact takes place. Consequently we do not want to touch the surface until the airplane is at its slowest possible flying speed. Very often if a landing site is soft it probably is also short and a combination of techniques is required. But in the simplest form, the soft-field landing assumes that you do have plenty of runway and no large obstructions to fly in over.

The flaps setting for soft landing will depend on the airplane and conditions, but remember a flatter approach means an easier transition in the landing flare. Lesser amounts of flaps, as long as the speed is controlled, might be helpful. During the landing flare a small amount of power is used to prolong the touchdown in an effort to get an even slower touchdown speed. The perfect soft-field landing would flare in parallel with the ground at two feet AGL. As the airplane coasts and floats through the flare, the airspeed will decrease. For every knot of lost airspeed there will be a corresponding reduction in lift, so the airplane will sink. When the sink begins, the pilot should add just a little control wheel back pressure to replace the lost lift, reduce the airspeed even more, and prolong the touchdown. This sink/back pressure response takes place two or three times during the length of the flare. Finally, the last addition of back pressure cannot replace enough lift because the airplane is simply too slow and it touches down. The airplane lands at the slowest possible speed with the pilot holding the wheel all the way back. Now the main wheels are on the ground, but the job is not done. We don't want the nose wheel to drop in a hole and flip us on our back, so we hold the back pressure and together with the small amount of power that we left in, the nose wheel lands last and slowest. That would be the perfect technique, but some problems can arise.

The soft-field technique's prolonged flare means that the airplane remains in the air longer and at an ever-decreasing speed. During that flare a crosswind can be a big problem. The airplane is only 2 feet above the surface, which is not enough room to correct for wind by placing one wing low and holding with rudder. Soft-field landings are susceptible to drift. You should not even attempt the soft-field landing with strong, gusty crosswinds. Also, for every second that you remain in the air with diminishing airspeed, the more hazardous a go-around would become. And this technique of an ever-slower flare will only result in a soft landing if the airplane is just above the surface when the flare takes place. If the speed dissipates during an extended flare and the airplane is still 10 feet in the air, it will drop the remaining distance. Anything that drops from 10 feet or greater could not be considered "soft."

The Practical Test Standards also addresses the short-field landing. The FAA standards says, *"[the pilot should] make smooth, timely, and correct control applications during the roundout (flare) and touchdown."* The mention of "timely and correct" refers to the pilot's application of back pressure as the airplane begins to sink so that the flare is prolonged until the slowest flying speed is reached at touchdown. This standard is the same for both private and commercial pilots.

Crosswind Landing

The crosswind landing is one of the most challenging maneuvers in all of flying. I learned to fly at an airport with only one air strip, so I got plenty of crosswind practice because there was really no choice. I had several flights canceled just because the wind was strong and there was no alternate runway that could be used to cut down the crosswind component. Pilots who fly from airports with multiple runways, and therefore have a choice, often have crosswind problems when they arrive at an airport with no choice. The crosswind approach has two basic techniques: the slip and the crab. The crab or wind correction angle is the method we use to overcome wind drift in all phases of flight except takeoff and landing. One of the basic lessons of flight is that airplanes do not always travel in the direction that you point them. The pilot might tightly hold a heading of north but a strong wind from the west will push the airplane northeast. The nose is on 360° but the direction of travel is 030°. To counteract the drift the pilot can fly a crab angle, something like 330° so that the actual course across the ground stays at the desired 360°. This works great in the air, but you cannot use it to land because when the nose is turned into the wind the wheels also turn into the wind. You should not allow the wheels to touch down when they are not aiming in the direction of travel. When this does happen a "side load" is placed on the tires that can lead to loss of control.

So a crab angle can be used during most of the traffic pattern and even into the final approach, but eventually you must leave the crab angle and get the wheels aligned with the runway. But if you simply come out of the crab angle without any further correction, the airplane will drift downwind again, and that can cause a loss of control.

I prefer the slip method because there will be less to do in the flare and touchdown. I recommend that you roll out on final approach and instead of lining up with the runway centerline, line up with the upwind edge of the runway. Then see how long it takes for the wind to blow you over to the centerline. The rate at which this happens will tell you just how much slip correction will be required. When you arrive in alignment with the centerline, roll the upwind wing down. Now ordinarily when banking the airplane, a turn in the direction of the bank will begin. In this case you don't want to turn away from the runway so you add opposite (downwind) rudder to stop the aileron from turning the airplane. Now it becomes a balancing act. The aileron wants to take you one way; the rudder wants to take you the other way. Apply the aileron and rudder pressure so that they cancel each other out. The result will be a wing-low yet runway-aligned approach. As you get closer to the surface, expect the wind to change, shift, or become more turbulent. This requires instantaneous reaction on the rudder and aileron to correct for the changes. Make smooth control pressure changes so that the upwind wing remains lower than the downwind wing. The wings look banked during this time, but the wheels are lined up straight down the runway. Continue holding the wing-low attitude right into the flare. With the upwind wing down, the upwind main wheel will also be down. Touch down on the upwind wheel first, while rudder continues to hold the wheels straight with the centerline. The downwind wheel lands second, and the nose wheel last. When the downwind wheel touches, the wings will no longer be banked and therefore the airplane no longer has wind correction. Here is where the airplane could still be carried downwind off the centerline or even off the runway. To prevent this, after the wheels are on the ground, apply full aileron into the wind. This has the effect of sealing the airplane to the ground with the wind helping hold the airplane to the surface.

The slip crosswind method provides no side-load touchdowns and good directional control in the air and during the rollout, but it does require finesse and practice.

The Private Pilot Practical Test Standard for normal and crosswind landing requires: "[the applicant must] *Touch down smoothly at the approximate stalling speed, at or within 400 feet beyond a specified point...*" Commercial pilots have 200 feet.

Killing Zone Survivor Story

My time in the "zone" was a rather uneventful one, except for one instance. I had accumulated about 124 hours total time and thought I was a pretty good VFR pilot. This particular flight was one where I would not only build total time; I would also satisfy the long cross-country requirement for my commercial certificate.

I was flying a Cessna 152, and I remember it was equipped with long-range tanks. Having flown from Kinston, North Carolina to Roxboro where I would pick up my wife for a day trip to Hilton Head, South Carolina, I felt confident about the trip. Despite the ever-present forecast of thunderstorms in the middle of the summer, the weather was beautiful. My wife had not flown with me much so I wanted to make a good impression. I pointed out the visual checkpoints along the way. She was duly impressed when the island came into view.

I could tell that the wind was a noticeably steady, onshore breeze. The runway at Hilton Head runs relatively parallel to the shoreline so this presented a cross wind from my right as we were advised to land on runway 3. The runway is lined on both sides by tall pine trees. Even though I had set up the landing well, tracking the centerline the whole time using the side-slip method, I did not count on the trees as being such a good windbreak. Right about 20 feet above the ground, I suddenly lost all my wind. I also lost my concentration. This led to my having much too much cross-wind correction for the actual winds at runway level and I was not quick enough, nor experienced enough, to compensate.

We touched down more sideways than I ever had before that day, or ever since. It was ugly. I never had heard such a squalling of the tires, nor been thrown sideways in the seat. As I looked up I could see that we were heading for the runway lights off to the right, so I kicked hard left rudder and simultaneously mashed the brakes. The tires chattered with protest, but we stopped without

hitting anything. Both of us shaken from the experience, we taxied to the ramp where someone asked us if the crosswind was bad. I just smiled sheepishly and said, "yeah."

This survivor story was submitted by Phil Heitman. Phil was a former student, but today flies for a regional air carrier and has written several books of his own. His experience was not significantly different from some of the sample accident reports. Phil did not lose control to the point where the airplane ran off the side of the runway, but others were not as fortunate.

Cross Control

There is a landing setup that pilots can inadvertently find themselves in that can be very dangerous. It takes place when there is a crosswind, but the danger exists prior to reaching the runway. The biggest danger zone is during the base to final approach turn when the pilot has not anticipated the effects of the crosswind.

When you approach an uncontrolled airport, you must always resolve the issue of which way to land. You could fly over the airport and look down at the wind sock, but that should be the last resort. You should be able to determine the active runway farther out, allowing time to maneuver for the proper traffic pattern entry. You should already know the wind's general direction and runway alignment so you already have a good idea which runway to use. But many uncontrolled airports have Automated Weather Observing Systems (AWOS) or Automated Surface Observing Systems (ASOS) that transmit a local weather broadcast including the prevailing wind. The frequency to hear the broadcast is written on the sectional chart with the airport's information. If there is no AWOS or ASOS, then tune in the Common Traffic Advisory Frequency (CTAF), which is also listed on the chart. There may be other airplanes already in a traffic pattern. You will want to join the flow of traffic so that the question of which runway is answered for you by the other pilots.

What if there are no other pilots and the wind is a direct crosswind over the airport's single runway. Which way do you land in that case? There will be a crosswind no matter what, so is there a preference? Yes. If you have the choice you should select the runway that provides a headwind on the base leg. If you choose the opposite runway, you would then have a tailwind on the base leg and that can lead to the cross control situation as you turn from base to final.

Figure 6.5 illustrates the traffic pattern that sets up a tailwind on the base. Many accidents have taken place because the pilot does not properly anticipate the turn to final. Under normal circumstances, without a crosswind, there would not be a tailwind on the base leg. Pilots get accustomed to the time it takes to fly the base and usually there are several prelanding chores to accomplish on base: flap settings, power reductions, etc. With a tailwind on base the airplane's ground speed on that leg will be much faster than normal. This means that the airplane is quickly carried through the base and the need to turn final comes up sooner than expected. The wind carries the unexpecting pilot across the extended runway centerline. The airplane overshoots final. The pilot now sees what has happened and attempts to make up for lost time. The pilot instinctively starts a steep turn toward the runway, but when the back angle exceeds 45 degrees, the normally safety-concious pilot remembers that steep turns in the traffic pattern are not safe, so the pilot shallows the bank slightly. Then it becomes apparent that with the reduced angle of bank the airplane will never get around the turn in time, so the pilot "helps" bring the nose around by applying rudder to the inside of the turn. This points the nose at the runway, even though the bank is not too steep. It looks as though the problem is solved. But wait; the ball in the inclinometer has crept out of the center position. In Figure 6.5, the pilot is holding right aileron so that the bank will not get too great and simultaneously holding left rudder to continue the turn to the runway. The airspeed is dropping all during this time. The airplane is near a turning, accelerated, uncoordinated, low-altitude stall. If the stall does occur here, the airplane will be virtually unrecoverable in the altitude remaining. If the traffic pattern had been flown with a headwind on the base leg, then all this could have been avoided. If this setup does occur, and you find yourself in an overshoot of final approach, the best solution is to make a go-around. If you cannot switch runways because of other traffic, then do a better job of anticipating the final turn during the second attempt.

Go-Around

On each and every approach to land, you must not feel rushed, you must be confident about the conditions, and it must feel right. If it just

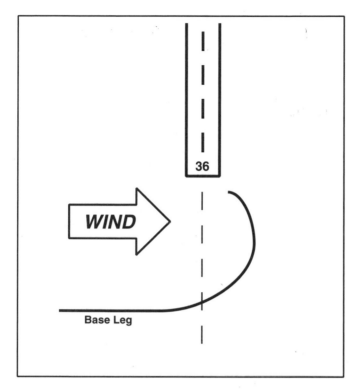

Fig. 6.5 A tailwind on the base leg can cause a runway centerline overshoot.

doesn't feel right or look right, then you should make a rejected approach or a go-around and set it up again so that it does feel right. But the go-around itself can be hazardous. It is very important to know what is suggested by the aircraft manufacturer for the go-around procedure. Many things are happening at once during the go-around, and it is easy to get distracted. The airplane must be transitioned from a landing configuration with all its drag to a climb configuration with reduced drag. This is taking place at the same time that the airplane's direction of travel is radically being switched from a descent and low power to a climb at high power. Typically, flaps and landing gear are coming up, torque and p-factor are increasing, and the pilot is also watching for traffic and maintaining directional control. The pilot's workload changes from a prepared, managed, undercontrol approach to a quick decision followed by an unplanned reversal of power, direction, and configuration. It is like juggling three balls and then unexpectedly someone throws you three new balls. You have to suddenly work the new balls into the act. It would be very easy to drop at least one ball.

NTSB Number: BFO93FA138. Clearfield, Pennsylvania

The airplane was observed to enter the traffic pattern after the pilot requested and received an airport advisory. The pilot did not report any problems, and the engine "sounded good" during the time witnesses observed the airplane turn onto a "very short final approach." As it crossed the threshold about 20 feet above the runway, it "ballooned" and "yawed." Full engine power was heard, the airplane turned right as it climbed to about 100 feet and headed towards some hangars. It steepened its bank to the right, pitched down, and impacted the ground.

Probable Cause:

The pilot's failure to maintain airspeed during the go-around.

After overflaring for the amount of airspeed, and getting some sideways yaw, the pilot did not like the landing setup and attempted the go-around. The decision to go around was probably a correct one. It did not look right to the pilot. The witnesses reported a very short final, so it is quite possible the landing was rushed and that accounted for the extra speed and the ballooning. At this point the pilot was uncomfortable and decided to try again. The power came up to full, but with everything happening at once, the airspeed now got too slow. Directional and flight control was lost and the 115-hour private pilot was killed. The accident sequence started on the downwind leg when the airplane and the pilot were not getting prepared to land. One problem led to another, and a tragedy was the result.

Landings provide the pilot with a constant challenge. Each one is new with different variables of wind, surface condition, altitude, sun, traffic, turbulence, and obstructions to deal with. Successful, uneventful landings are the product of good planning, good judgment, and being familiar with the airplane. The regulations require that pilots make three takeoffs and three landings every 90 days in order to carry passengers. But the regulations do not dictate the conditions that influence these landings. You could have made your three takeoffs and three landings where there was little wind, on a dry, hard surface runway that was

8000 feet long. And then later face a landing uphill, with a crosswind, on a 3200-foot runway with trees at both ends. You are legal to make the landing, but are you prepared? Using good judgment means getting some instruction and practice in adverse conditions. It also means occasionally diverting to another airport where the odds of a safe landing are better.

CHAPTER SEVEN

Runway Incursion

NOT UNTIL RECENTLY have on-the-ground accidents been a focus of attention. But while other areas of accidents have declined, runway incursions have gone up. From 1993 to 2000 there was a 73% increase in reported runway incursions. Maybe the problem was always bad but the system of reporting improved, revealing the true level of the problem. The FAA defines a runway incursion as, "Any occurrence at an airport involving an aircraft, vehicle, person, or object on the ground that creates a collision hazard or results in a loss of separation with an aircraft taking off, intending to take off, landing, or intending to land." The FAA further identifies the problem by defining the word "occurrence" as it appears in the runway incursion definition. An occurrence is:

1. A pilot deviation that is a violation of the Federal Aviation Regulations.
2. A controller deviation or operational error that violates minimum separation between aircraft or aircraft and obstructions.
3. A vehicle or pedestrian deviation that violates movement areas without ATC authorization.

The worst accident in aviation history was a runway incursion. It took place in 1977 at Tenerife in the Canary Islands. Two Boeing 747s collided on the ground and 583 people were killed. The visibility was low due to fog, and there was construction on the airport that made the normal taxi route to the runway impossible. One 747 began a takeoff roll, while the other 747 was on the runway.

The Northwest & Northwest Collision

NTSB Number: DCA91MA010A. Romulus, Michigan

On December 3, 1990, at 1345 EST, Northwest flight 1482, a DC9, and Northwest flight 299, a Boeing 727, collided near the intersection of runway 09/27 and 03C/21C in dense fog at Detroit Metropolitan/Wayne County Airport, in Romulus, Michigan. At the time of the collision, the B727 was on its takeoff roll, and the DC9 had just taxied onto the active runway. The B727 was substantially damaged, and the DC9 was destroyed. Eight of the 39 passengers and 4 crewmembers aboard the DC9 received fatal injuries. None of the 146 passengers and 10 crewmembers aboard the B727 were injured.

Probable Cause:

Lack of proper crew coordination, including virtual reversal of rolls by the DC9 pilots, which led to their failure to stop taxiing and alert ground control of their position uncertainty in a timely manner before and after intruding onto the active runway. Contributing to the accident were (1) deficiencies in the ATC service provided by Detroit tower, including failure of ground control to take timely action to alert the local controller to the possibility of a runway incursion, failure to use progressive taxi instructions in low-visibility conditions, and issuance of inappropriate and confusing taxi instructions compounded by inadequate backup supervision for the level of experience of the staff on duty; (2) Deficiencies in surface markings, signage, and lighting at airport and failure of FAA surveillance to detect or correct any of these deficiencies; (3) Failure of Northwest Airlines to provide adequate cockpit resource management training to line aircrews. Contributing to the fatalities was an inoperability of the DC9 internal tail cone release mechanism. Contributing to the number and severity of injuries was the failure of the crew of the DC9 to properly execute the passenger evacuation.

And then less than two months later…

The US Air and Skywest Collision

NTSB Number: DCA91MA018A. Los Angeles, California

Skywest flight 5569 had been cleared to runway 24L, at intersection 45, to "position and hold." The local controller, because of her preoccupation with another airplane, forgot she had placed Skywest 5569 on the runway and subsequently cleared US Air flight 1493, a Boeing 737, for landing on the same runway. After the collision the two airplanes slid off the runway and into an unoccupied fire station. The tower operating procedures did not require flight progress strips to be processed through the local ground control position. Because this strip was not present, the local controller misidentified an airplane and issued a landing clearance.

Probable Cause:

The failure of the Los Angeles air traffic facility management to implement procedures that provided redundancy comparable to the requirements contained in the National Operational Facility Standards and the failure of the FAA air traffic service to provide adequate policy direction and oversight to its air traffic control facility managers. These failures created an environment in the Los Angeles air traffic control tower that ultimately led to the failure of the local controller to maintain an awareness of the traffic situation, culminating in the inappropriate clearances and subsequent collision of the US Air and Skywest aircraft. Contributing to the cause of the accident was the failure of the FAA to provide effective quality assurance of the ATC system.

Even though there had been ground collisions before, it was these back-to-back accidents that put "runway incursion" in the spotlight. Most thought, if we can just keep these airplanes apart in the air, then the ground should be easy. We do not think that anymore and cannot take ground operation lightly.

Killing Zone Survivor Story

The US Air and Skywest collision was created by a controller error and the environment that surrounded that controller. This survivor story comes to us from an anonymous air traffic controller whose story is very similar.

NASA Number: 421891

I had just relieved the previous controller about one minute earlier and was trying to establish vertical separation between a previous aircraft departure and a VFR Cessna overflight [of the airport] before switching to aircraft over to departure [control]. I was also trying to turn up the radar scope's map and alpha numeric brightness which was very low when aircraft X called ready to go. I scanned the runway, but apparently not far enough on final. I cleared aircraft X for takeoff and resumed adjusting the scope when a short time later aircraft Y, a Lear on final approach who had been cleared to land by the previous controller, asked what aircraft X was doing. I looked up and saw the Lear (aircraft Y) on short final with aircraft X past the hold line with his nose over the runway edge. I told the Lear to go around. The Lear did not respond and continued to land. The Lear pilot apparently thought that aircraft X was at fault and said he caused a runway incursion. It was my fault. Had I set up the scope prior to taking the position, or simply told aircraft X to hold short while I was busy with the departing traffic and the VFR traffic, and taken an extra moment to fully settle into the position, this incident would not have occurred. I've heard that the first few minutes and the last few minutes on a radar position are when the most incidents happen in ATC, and now I know from experience. Be extra vigilant when taking the position and don't let the fact that there is very little traffic lull you into letting your guard down.

The pilot of the Lear Jet, described as aircraft Y saw the other airplane as it entered the same runway he was landing on. But this took place during daylight hours. The US Air pilots in Los Angeles did not have that advantage. This was a close call, one that points out that pilots must be vigilant and be ready to overcome controller mistakes.

The Tenerife, Los Angeles, and Detroit runway incursion accidents are rather famous (infamous). You probably remember these accidents from the news coverage when they took place. The pilots involved in those accidents were not inexperienced. Do general aviation pilots with less experience get involved in runway incursion accidents? Unfortunately the answer is yes and they occur at both controlled and uncontrolled airports.

NTSB Number: MIA00FA103A. Sarasota, Florida

On March 9, 2000, about 1035 eastern standard time, a Cessna 172K on a personal flight, and a Cessna 152 on an instructional flight collided during takeoff on runway 14 at the Sarasota-Bradenton International Airport, in Sarasota, Florida. Visual meteorological conditions prevailed at the time and no flight plan was filed for either flight. Both aircraft were destroyed. The airline transport-rated pilot and another pilot in the Cessna 172, and both the flight instructor and student pilot in the Cessna 152, all were fatally injured. Both flights were originating at the time of the accident. Air traffic controllers at the FAA Sarasota-Bradenton International Airport Air Traffic Control Tower stated that the flight instructor of the Cessna 152 called for taxi instructions from the Dolphin Aviation ramp and was told by the ground controller to taxi to the end of runway 14. The ground controller then went on a rest break and a supervisor took over the ground control position. Draft transcripts of communications show that about 2 minutes after the Cessna 152 called, the pilot of the Cessna 172 called for taxi instructions from the Jones Aviation ramp and was told by the supervisor ground controller to taxi to runway 14. [The Dolphin and Jones FBO ramps are on opposite sides of runway 14; See Figure 7.1.] *The supervisor stated that he thought the Cessna 172 was leaving from the Dolphin Aviation ramp and would taxi to the end of runway 14 and not the intersection of runway 14 and the foxtrot taxiway, which leads from the Jones ramp. The paper strip [air traffic controllers record of the movements of the Cessna 172] was marked for the end of runway 14. Draft transcripts show*

that the pilot of the Cessna 152 called ready for takeoff at 10:30:46.
The local controller told the pilot to hold short of the runway. At
10:32:51, the pilot of the Cessna 172 called the local controller stat-
ing he was the number 2 aircraft ready for takeoff. The pilot was told
to hold short of the runway. At 10:34:00, the pilot of the Cessna 152
was told by the local controller to taxi into position and hold. The
local controller then cleared a third aircraft into position and hold at
the intersection of runway 14 and the foxtrot taxiway. That aircraft
was then cleared for takeoff. At 10:34:54, the local controller cleared
the Cessna 152 for takeoff on runway 14. The local controller stated
that he then saw the paper strip for the Cessna 172, which showed
that the aircraft was at the end of runway 14, looked out and saw a
high wing Cessna in the first position at the end of the runway, and
thought it was the 172. At 10:34:57, he then cleared the 172 into
position and hold on runway 14. He then diverted his attention inside
the tower cab, and a short time later he observed the fire which
resulted from the collision of the Cessna 172 and Cessna 152. The
sound of an emergency locator transmitter was heard at 10:35:13. A
witness stated he observed the 152 begin a takeoff roll from the end
of runway 14. At about the point where the aircraft obtained takeoff
speed, the Cessna 172 entered the runway from a taxiway on the left
side of the runway [taxiway foxtrot]. The Cessna 152 lifted off and
got a few feet in the air and initiated a right turn, in what appeared
to be an attempt to avoid collision with the 172. The Cessna 152 then
appeared to stall and crash into the 172. Both aircraft immediately
erupted in flames and came to rest on the runway together.

The supervisor ground controller entered the picture halfway into
the sequence and believed that the Cessna 172 was behind the Cessna
152 on the same taxiway. The controller did not put taxiway foxtrot on
the information strip as is standard procedure. Maybe if the controller
had written down foxtrot on the strip it would have alerted him to the
error. Information strips on previous airplanes departing Jones on taxiway foxtrot that day did have the notation: "foxtrot." Forgetting to
write down the taxiway on the strip did not by itself cause this accident,
but it was one link in the chain.

There were three airplanes involved. Two were at the intersection of
foxtrot and runway 14 (see Figure 7.1) and one was at the end of runway
14. But the controller had it backwards. He thought one was at the inter-

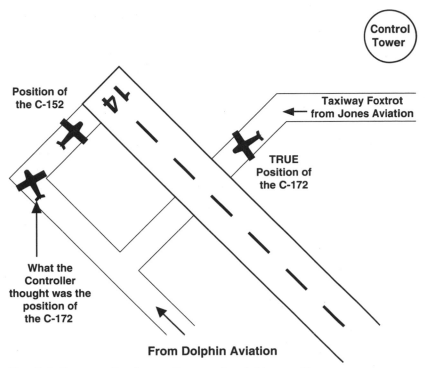

Fig. 7.1 Sarasota-Bradenton International Airport diagram.

section and two were at the end. At 10:32:51, the pilot of the Cessna 172, who was actually number 2 at the intersection, called the local controller stating he was "the number 2 aircraft ready for takeoff." The controller must have thought he meant number 2 behind the Cessna 152 at the end of runway 14. The Cessna 172 pilot was not as clear and not as aware as he should have been here. The Cessna 172 pilot should have noticed that another airplane, the Cessna 152, was also preparing for takeoff down at the end of the runway. If he had seen the other airplane, he could have made a better radio call by saying, "ready for takeoff at the intersection of runway 14" or "ready for takeoff runway 14 at foxtrot." Either way this would have clarified the Cessna 172's exact position. The 172 pilot, however, made a vague statement that allowed the controller to continue to believe the wrong position. The first airplane at the intersection was given permission to taxi into position and hold, and the Cessna 152 was given position and hold behind him. Those two airplanes were not staggered in position on runway 14. The lead airplane was given the takeoff clearance and away they went. The Cessna 152 was given a takeoff clearance and began the takeoff roll. The controller then, still thinking that the 172 was behind the 152, gave the 172 the clearance to roll onto the runway and hold. But the 172 being at the intersection, not the end, rolled in front of, not behind, the 152. Figure 7.1 illustrates the approximate location of the control tower at Sarasota relative to the end of runway 14. When the controller looks to the end of runway 14, the line of sight is directly over the intersection where the Cessna 172 was waiting. The controller would have had to look over the 172 to see the end of runway 14. Why didn't the controller see him? Did the controller, because of his belief that the 172 was on the other side, simply disregard anything he saw on the near side?

And the biggest question and the best possible prevention of the accident: Why did the Cessna 172 pilot pull out in front of an oncoming airplane on takeoff roll? The 172 pilot was the closest to the 152 and would have had an unobstructed, broad daylight, view back down the runway. But the pilot of the 172 also must have held a belief that the runway was clear just because the controller said it was clear. This tragedy can be so clearly broken down after the fact, but nobody saw the chain forming during the fact.

NTSB Number: BFO94FA92B. Millville, New Jersey

According to the Flight Service Station (FSS) frequency transcription, a Cessna at the hold short line for the active runway called for takeoff clearance. The airport was uncontrolled but had an FSS on the field. The FSS personnel asked if he needed an airport advisory and the person stated "no." The FSS personnel states, "...advise when airborne." The Cessna pilot broadcasted that he was "...taking the active." The Cessna was taxiing in a right turn near the centerline of the runway when it collided with the Piper that had just landed and was starting to rotate for takeoff. The Piper's propeller blades impacted the aft outboard section of the Cessna's left wing. The pilot of the Piper stated that he did not hear the Cessna pilot's last broadcast and did not see the Cessna until it was too late for evasive action. The FSS transcription revealed that the last transmission made by the Piper pilot was that he was on the downwind leg of the traffic pattern and was inbound for a touch and go. Both Cessna occupants were pilots. It was not determined where the

Cessna occupants were seated nor who was controlling the airplane at the time of the accident.
   Probable Cause:
   The failure of the Cessna's pilot in command to perform an adequate visual lookout.

It was clear that the pilot of the Cessna was not familiar with the proper operations at an airport with an FSS. The Cessna pilot asked questions of the FSS as if the FSS were a control tower. Evidence of this is when the pilot asked for a takeoff clearance. Only a control tower can issue a takeoff and/or landing clearance. An FSS can only give advisories. A control tower controller is up high in an actual tower designed so that the controller has visibility to the airport. An FSS specialist is inside a flat building that may not have an outside view of the airport. The FSS may not even be directly on the airport grounds.

When an FSS is in operation at an otherwise uncontrolled airport, the airspace within 10 statute miles of the airport becomes an Airport Advisory Area. There used to be many of these, but today there are only a few. Most of the outlying Flight Service Stations were decommissioned in the 1990s and consolidated into the larger Automated FSS. Now there may be only one FSS in your entire state. So Airport Advisory Areas are rare and that may have contributed to this accident pilot's lack of familiarity.

The FSS can only give advice, hence the name "advisory area," and does not authorize operations like a control tower. Communications with the FSS in such an area is even voluntary, but highly recommended. When the FSS specialist asked if the Cessna pilot wanted an advisory, the pilot responded, "No." The pilot must not have realized that an advisory would have informed him of other traffic in the area. When the FSS said, "...advise when airborne," the Cessna pilot must have misunderstood this to mean, "cleared for takeoff." But the FSS specialist inside his office could not have known if the runway was clear or not. The Cessna pilot, thinking he had just been told that the runway was clear and belonged to him alone, called out "...taking the active" and without looking pulled out in front of the Piper, who was in the middle of a touch-and-go. The Cessna pilot violated one of the rules we learned earliest in life: Look both ways.

NTSB Number ATL97LA004A. Angier, North Carolina

An Aeronca 65, had completed a full stop landing and was in the process of back taxiing on the 100-foot wide runway when it collided with an American AA-1, which was on a takeoff roll on the same runway. The collision occurred after the Aeronca pilot executed a right turn in an attempt to avoid a head-on collision. The American AA-1 collided with the left side of the Aeronca. At this airport, the common practice is to back-taxi on the side of the active runway. Both pilots reported that the runway conditions were excellent, and the sod surface was dry.

Probable Cause:

The Aeronca 65's pilot selection of the wrong taxi route, and his inadequate visual lookout during the back-taxi.

Even at airports with nonhard surface runways, an incursion can take place. At this airport there were no clearly defined taxiways; the local procedure was to turn around and return to the end of the runway along the side. This time he did not turn all the way off the designated runway. There were four people in the two airplanes. Nobody was killed, but there was one serious injury because neither pilot was watching.

These incursion examples took place on runways themselves, but accidents have happened and people have been hurt and killed in taxiway incursions.

NTSB Number: LAX86LA284. Los Angeles, California

The private pilot had no prior experience at the Los Angeles International Airport and the pilot was unknowledgeable of the existence of the airport motor vehicle traffic service roads. The pilot was given an ATC clearance to taxi, and in the darkness the pilot mistook a service road for the desired taxiway. Also, the pilot reported to the NTSB that he was in a hurry to fly from LAX to San Diego because he had to return to work. The pilot acknowledged that, while taxiing, he had momentarily diverted his attention to inside the cockpit. When he looked back outside he observed a truck, but he believed his right wing would clear it. The truck, which had stopped on the service road and was holding for the airplane, was struck on its left rear side by the Cessna's right wing.

The pilot was unhurt, but the Cessna 172 sustained substantial damage and the truck driver suffered a minor injury when he leaped from the truck.

NTSB Number: SEA94LA130A. Jackson Hole, Wyoming

While taxiing for takeoff, the pilot of a Christen Eagle II, became distracted when he looked into the cockpit to check the adjustment of the mixture control. When he looked up, it was too late to keep from hitting a Cessna 172, that was parked in the run-up area preparing for departure. The Eagle pilot attempted to turn his aircraft to avoid the collision, but both right wings impacted the 172.

Probable Cause:

The failure of the pilot of the Cristen Eagle II, to see and avoid the Cessna 172, and the pilot of the Eagle's diverting his attention from the operation of his aircraft.

In flight we can quickly look away to inside instruments and controls, and we even look side to side instead of just ahead. But during taxi operations we must operate the airplane more like a car and constantly watch where we are going. Both airplanes were substantially damaged, but only minor injuries were sustained by the two people in the 172.

NTSB Number: LAX84LA230A. Ramona, California

The pilot of a North American T-6G, had just landed and was taxiing down a parallel taxiway to the approach end of the runway to take off again. At about the same time, a Cessna 150 had been stopped at the intersection of the parallel taxiway and another adjoining taxiway. The Cessna 150 had stopped so that the flight instructor could get out of the airplane so that the student could proceed with his third supervised solo flight. As the T-6 approached the intersection where the Cessna was parked, the T-6 made a shallow S-turn to the left, then an S-turn back to the right. At that

time, the left wing of the T-6 impacted the empennage of the Cessna from the left rear. The flight instructor had just deplaned and was standing on the ground next to the Cessna. He was knocked off balance and grabbed the strut to keep from falling toward the propeller. He then fell under the Cessna and injured his knee. None of the pilots were aware that the collision was about to occur.

The collision caused minor damage to the airplanes, but probably scared the flight instructor to near death. My very first flight instructor told me about another CFI who got out of an airplane so that a student could solo, and promptly walked into the propeller. It was his rule, and it became mine, never to open any airplane doors with an engine running.

Ground Operations Guidelines

Whether on a taxiway or near a runway, we must be aware of where we are and what is going on around us. Use the following guidelines when operating at any airport, and future accident chains can be broken in time.

1. Know the runway and taxiway markings. You must know what hold lines mean, and where it is approved to taxi. Any taxiway that has the blue taxiway lights alongside is called a "moving area," and ground control authorization is required to taxi on that part. Know the markings for an ILS sensitive area. Understand the meaning of airport lights and colors.

2. Always have a map of airport layouts with you. The VFR sectional charts do not include airport layout maps, so I always get a copy of the map from an instrument chart. Instrument chart makers do include detailed airport layouts. I always make a copy of the layout and tape it to my student's navigation record when they depart for a larger airport. Taxiways are lettered, not numbered, so taxi instructions can be confusing if you do not have a map: "taxi south on Kilo, then left on November, but hold short of Romeo." Be aware of airport construction and NOTAMs of closed taxiways and runways.

3. When on the surface of any airport ask for "progressive taxi" whenever you are not completely sure of your position or how to

get to your destination. Even at an uncontrolled airport, make a call on unicom if you are not positive.

4. Give every detail you can when you talk. When you first radio to ground control after landing, don't say, "I'm clear the active" because there may be four active runways and they will not immediately know which one you mean. Instead say, "Ground, 1234A is clear runway 20R at Tango Six." This precisely tells the controller which runway you have come off of and at what location you came off. When ready for takeoff say: "Tower this 1234A ready for takeoff at the end of runway 9." or you could say, "Tower 1234A is ready for takeoff, number 2 in sequence at runway 32, intersection Charlie." Be very descriptive.

5. Turn on lights day or night.

6. Taxi into position and hold *slowly*. Sometimes when traffic load is high, a controller will assign the pilot to "taxi into position and hold." This means: get ready, get set, but don't go yet. The controller at that moment cannot issue the takeoff clearance because the runway is not clear yet. Either an airplane has just landed or taken off on your runway or a crossing runway. The controller is buying time. They can't let you go yet, but they will want you to go on a moment's notice. This is good time management for the controller, but it may not be in the pilot's best interest. When an airplane is in position on the runway, the pilot's back is turned to any airplanes that might be approaching that runway. You literally become a sitting duck. How can this risk be reduced? When controllers tell you to "position and hold" they do not tell you how fast to get into position. Therefore roll out slowly. Cross the hold line and enter the runway at a snail's pace. Positioning slowly will allow you to look back in the direction of oncoming traffic longer. Done properly, the slow taxi time will equal the takeoff delay and the controller will issue the takeoff clearance without your ever having to completely turn your back on other airplanes. Keep all delays on the runway to an absolute minimum. Position and hold is not a good place to complete checklists or carry on conversations.

7. Enter a taxiway or runway just like you would cross a street—look both ways! Do not assume that everyone else is operating properly. Drive your airplane on the ground defensively.

8. If you ever find yourself turned around on the airport surface, or confused about where you are and what you are supposed to do: STOP and ASK before moving any further.

Killing Zone Survivor Story

A pilot and air traffic controller tell their side to the same story.

The Pilot's Side of the Story: NASA Number 410622

Upon taxiing out from Charlotte, North Carolina, I picked up my IFR clearance from the tower and taxi instructions. The controller told me to taxi to runway 18L at intersection Alpha, which I did. The controller cleared me for takeoff from runway 18L at intersection Alpha and to fly a heading of 210 degrees, which I did. Later I was informed by the Charlotte approach control by telephone that runway 18L had been closed, and that I came close to hitting a vehicle that was on the runway.

The Controller's Side of the Story: NASA Number 410925

Airport operations closed runway 18L for maintenance. I turned off the runway lights and put the closure on the ATIS. Twenty minutes later I (working alone in the tower) taxied an aircraft to the closed runway and issued a takeoff clearance forgetting that the runway was closed. The aircraft departed over vehicles on the runway. Factors to include: (1) airport operations—when closing a runway and having vehicles operate on the runway, should have better illuminated vehicles. (2) Pilot—the pilot did not listen to the ATIS or question why the runway lights were turned off. He also did not see the vehicle. (3) FAA—although the FAA knows runway incursions to be a problem they had no memory aids to alert the controller that the runway was closed.

This ground near-miss could have been a fatal accident for the pilot and the vehicle operator, and there are plenty of questions that could be asked. Why would a pilot take off from a Class B airport like Charlotte at night without runway lights? That should have been very suspicious.

And the controller seemed to blame a lack of vehicle lights, the pilot, and lack of memory aids instead of himself. As pilots we should remember this story whenever something does not seem right. Never be afraid to ask the controller about something you are not sure of. "Hey, are the runway lights on 18L inoperative tonight?" If the pilot had asked that obvious question, the threat would have been eliminated.

Always remember that the trip actually begins when the aircraft first moves. All too often pilots get into a flight-only mindset. This means that they take ground operations for granted, mentally preparing for the flight portion only. When basketball players take the court, they really do not turn on their full concentration until the ball is tipped and the clock is started. Sometimes pilots are like that. We do not turn on our full concentration until takeoff and the in-the-air part of the trip begins. The sample runway incursion accidents of this chapter should remind us that we must be at full awareness the moment we take the court, which begins the moment we walk up to the airplane.

CHAPTER EIGHT

Midair Collision

MIDAIR COLLISIONS CONTINUE to be a problem, especially around airports. It makes sense that the greatest number of midair collisions would take place where airplanes converge: the airport. Eighty percent of midair collisions occur within 10 miles of an airport. And of those accidents, 78% occur at nontowered airports. In the 1990s there was an average of 26 midair collisions involving general aviation aircraft per year. These were not all fatal. The average number of fatal GA midair collisions was just short of 9 per year. The average number of people killed in these fatal accidents was 24 each year through the 1990s.

The first four midair accident examples that follow all happened near an airport or in an airport traffic pattern. They all took place during VFR weather conditions in the daylight. The visibility was reported between 7 and 20 miles in each case.

NTSB Number: MIA90FA67A. Sebastian, Florida

A midair collision occurred between a Piper PA-28 and an American AA-1A, as the two aircraft were on final approach to land on runway 13 at an uncontrolled airport. Both aircraft then crashed to the ground. The pilot of the AA-1A survived the accident, but due to injuries he could not recall details of the occurrence. The AA-1A was initially in left traffic behind the PA-28. A flight instructor (in traffic ahead of both aircraft) heard the PA-28 pilot report on a $1^3/_4$ mile final approach. He then heard the AA-1A pilot report on base leg. Thereafter, the instructor heard the AA-1A pilot report turning on a short final approach and became aware of the potential traffic conflict. The instructor then saw the AA-1A in a left turn on the right side of the final approach path, as if the pilot had overshot the turn to final. Also, he saw the PA-28 on final approach. Subsequently, the two aircraft converged and collided. During impact, the propeller and nose wheel of the AA-1A hit the empennage of the PA-28.

Probable Cause:

Failure of the AA-1A pilot to see-and-avoid the piper PA-28 which had entered traffic first. A factor related to the accident was: failure of the PA-28 pilot to see the American AA-1A, when the AA-1A was on a base leg and the PA-28 was further out on final approach.

Both of these pilots were flying alone, and sadly both were killed. Eighty percent of traffic pattern midairs happen on final when the pilot's complete concentration is on the upcoming landing. When we drive our cars we get accustomed to a "lane" mentality. This means that when driving a car, we figure that as long as we stay in our lane, then nobody will hit us. We drive looking ahead, with blinders on to anything else. In an airplane we cannot afford to fly with a "lane" mentality. We cannot become so focused on what is right in front of us. There are no lanes in the sky; another airplane could come at us from any direction: left, right, up, down, in front, or behind. Especially on final approach we must look in every direction while at the same time preparing for touchdown.

NTSB Number MIA95FA224B. New Smyrna Beach, Florida

The Aerospatiale TB-9, was observed on short final approach, and the Piper PA-38 had turned from base to final just above and behind the TB-9. Two pilots on the ground transmitted warnings to the aircraft but no action was taken. The aircraft collided; the TB-9 sustained stabilizer damage, nosed down, and crashed. The Piper landed without further incident. Pilots on the ground reported seeing the Piper performing takeoffs and landings, and heard the pilot making position reports. The pilots only observed

the TB-9 while on short final, and did not recall hearing any position reports from the pilots. They stated that there were many aircraft with similar call signs and they might have missed the calls. The operator of the TB-9 teaches his or her pilots to fly a 1.6 nm final approach at a 3-degree descent angle while making visual approaches. Other pilots stated that this practice conflicts with pilots who fly normal close-in approaches with a $^3/_4$ to 1 nm final approach leg. The aircraft flying the long final are at a lower altitude where a pilot making a normal visual approach would not expect to see conflicting traffic.

Probable Cause:

The failure of the pilots of both aircraft to see and avoid each other.

There was no injury to the pilot in the Piper, but all three people in the TB-9 were killed, and this accident added more fuel to the debate.

As long as I have been flying, and surely a long time before that, there has been an argument about the traffic pattern. In the first example the AA-1A turning from base to final hit the PA-28 that was already on final. But the PA-28 pilot had reported turning *"on a $1^3/_4$ mile final approach."* Is $1^3/_4$ miles out on final so far out that the AA-1A pilot lost the PA-28 or even thought that he had departed the pattern? In the second example, *"The operator of the TB-9 teaches their pilots to fly a 1.6 nm final approach..."* and that this practice is in conflict with pilots who fly a *"normal"* traffic pattern with a *"$^3/_4$ to 1 nm final approach leg."* So who is right? The regulation sheds only a little light on the question. The Part 91 flight rules say this: [91.113 Right-of-way rules (g) landing] *"Aircraft while on final approach to land or while landing, have the right-of-way over other aircraft in flight or operating on the surface."* The regulation goes on to say, *"...when two or more aircraft are approaching an airport for the purpose of landing, the aircraft at the lower altitude has the right-of-way, but it shall not take advantage of this rule to cut in front of another which is on final approach to land or to overtake that aircraft."* You can read this regulation both ways. Using the previous accident example, the TB-9 pilot was on final and could have claimed right-of-way. But the PA-28 pilot could also claim right-of-way because the TB-9 had used an excessively long final and was using that to cut in front of other airplanes in the pattern. The argument goes on.

I believe that if I ever have an engine failure while in the traffic pattern from the downwind leg on, and I cannot land on the runway, then I am too far out to begin with. You really do not want to completely depend on the powerplant to get you to the runway every time. If I go so far out that I cannot glide back, my pattern was too big. Of course, there will be times when the traffic in the pattern will force an extended downwind, but if I have my choice, I stay in tight. When I am following another airplane in the pattern I use a "wings crossing" technique to determine how far back I should stay behind the lead airplane. I do not turn base leg until the wings of the airplane I am following cross my own. The lead airplane is on final and my wingtip passes his while I'm on downwind. When the wings pass I turn base. The pilot ahead should announce on final his or her intentions to make a full-stop landing, touch-and-go, or stop-and-go. If they forget to make that announcement, I ask them. This makes it possible to plan for plenty of room.

In the final analysis, who has the right-of-way? Both or neither, depending on how you look at it. I feel the best solution is common pilot courtesy. Pilots in a traffic pattern must work and talk together. It really doesn't matter who had the right-of-way in a fatal accident. It is no good to be dead right. There is another regulation that should supersede all others: 91.111 (a) *"No person may operate an aircraft so close to another aircraft as to create a collision hazard."*

NTSB Number: NYC93FA158B. Lincoln Park, New Jersey

A Cessna 172 was in right traffic for the 4th in a series of landings on runway 01, and was established on final when the collision occurred. A witness observed the other airplane, also a Cessna 172, on base leg, turning right from base to final, just prior to impact. The witness observed the second 172 hit the first on the right side. After impact the first 172 fell vertically to the ground and impacted the east side of the road, in a ravine about one-quarter mile south of where the second 172 hit the ground. After the midair collision, the pilot of the second 172 tried to make the runway, but could not maintain altitude, and elected to land in the trees. As the airplane impacted the trees, the right wing separated, and struck an automobile traveling south on the road.

Paint transfer and impact marks were on both Cessnas. There were two minor injuries in the automobile.

Probable Cause:

The failure of both pilots to maintain adequate visual lookout while in the airport traffic pattern, which resulted in an in-flight collision.

Both pilots, flying alone, were killed. The pilot of the first 172 had 138 flight hours. This accident brings up yet another traffic pattern argument: Must airplanes approaching an uncontrolled airport fly a prescribed traffic pattern? Another regulation, 91.126 (b), has this to say, "*Direction of turns. When approaching to land at an airport without an operating control tower…each pilot of an airplane must make all turns of that airplane to the left unless the airport displays approved light signals or visual markings indicating that turns should be made to the right, in which case the pilot must make all turns to the right.*" That seems to be clear that turns are to be made either left or right and therefore a traffic pattern is required. But wait! Like most regulations this can also be read two ways. I had an FAA inspector make the case that this regulation does not in fact require turns of a traffic pattern. The regulation says, "*each pilot of an airplane must make all turns of that airplane to the left*" but what if you are not making turns? What if you are making a straight-in approach? His interpretation of the regulations was that if you were going to make turns, then they needed to be left (unless shown to be right), but if you did not intend on making any turns then the rule did not apply.

So must you enter a standard downwind, base, and final, or is a 10-mile straight-in final allowed? The answer is yes. Both are allowed, but again common pilot courtesy should prevail. Very often I approach uncontrolled airports in a cabin-class twin-engine airplane, and often a straight-in approach would be the best plan for me and my passengers. But I listen as far out as I can for traffic in a pattern. I ask for traffic advisories in hopes of learning where anybody might be. If there are other airplanes in a traffic pattern, I plan to get in line with the rest of them and enter into the flow of the pattern. But on occasion, like the middle of the night, when I have heard or seen no other traffic, I do a straight-in and reduce the time and noise of a traffic pattern fly-over.

NTSB Number: ATL93FA82A. Statesboro, Georgia

A student pilot in an American AA-1A, was on an approach to land on runway 5 of the uncontrolled airport, as a commercial pilot in a Cessna 414 was on approach to land on runway 14. The two runways intersected near their approach ends. The two aircraft collided as they were about to touchdown at dusk over the intersection. Witnesses on the ground stated that they observed lights on the Cessna, but did not observe any lights on the American. Witnesses in other aircraft in the area stated that they heard the pilot of the American announce his position in the traffic pattern and landing intentions, but did not hear the pilot of the Cessna on the unicom frequency. The pilot of the Cessna stated he announced his intention to land on runway 14 over unicom frequency 123.0. The published unicom frequency for the Statesboro airport was 122.8.

Probable Cause:

Inadequate visual lookout by the pilots of both aircraft. Factors related to the accident were: failure of the pilot of the American to illuminate his aircraft navigation lights, and improper radio communications by the pilot of the Cessna by selecting the wrong unicom frequency to monitor and announce his landing intentions and position.

The AA-1A pilot was killed. One person on the Cessna was killed. The student in the AA-1A had 79 flight hours. The Common Traffic Advisory Frequency (CTAF) was invented to prevent just this type of accident. Not so many years ago, before CTAF was printed on sectional charts, there was a great deal of confusion about which frequency to use. Some airports have part-time control towers. Before CTAF, we would use the control tower frequency when the tower was open, but we would switch to unicom frequency when the tower was closed. This worked except you always had to remember when the tower was open and closed. Pilots traveling to other airports did not always know when the tower was closed and would make their first call up when everyone else had already switched over to unicom. The same problem existed at airports with part-time FSS. After several accidents and near accidents, enough was enough—we needed just one frequency that would be used regardless of the time of day. That frequency today is called the CTAF. The CTAF frequency is printed on sectional charts and instrument charts. No matter what, transmit on the CTAF while in the air and taxiing on the ground. Unicom frequencies today are just for asking about fuel, arranging for a tie-down, or getting someone to call you a cab.

NTSB Number: NYC89FA135B. Bradford, Vermont

A Cessna 172M, with a commercial pilot and a student pilot passenger, collided in flight with a Cessna 172RG (Retractable Gear), which was being operated by a private pilot and a designated FAA check pilot on a commercial pilot checkride. Witnesses reported the two aircraft converged and collided while on headings of 30 and 45 degrees from each other. No evasive action was evident. During impact, the right horizontal stabilizer of the 172RG became imbedded in the windshield of the 172M, and the 172's propeller severed the tail of the 172RG. The 172RG went out of control and crashed. The occupants of the 172 made an emergency landing on an interstate highway with only minor injuries to themselves and minor damage to the aircraft.

Probable Cause:

Failure of the rated pilots in both aircraft to see and avoid a midair collision.

This accident did not take place in the immediate vicinity of an airport. The two airplanes collided while maneuvering, the 172RG pilot on his commercial checkride. The visibility was better than 10 miles. The pilots had medical certificates that had no waivers or limitations. The pilots could see well, but they did not give themselves the opportunity to use their good vision.

Pilots must understand the see-and-avoid concept. Anytime we are not in the clouds, we are responsible for our own separation. Even if you are on an IFR clearance, anytime you are not inside a cloud you must keep a vigilance outside the aircraft for other traffic. Air traffic controllers are almost powerless to help up. Controllers see aircraft targets on their radar displays everyday on collision courses. The two aircraft that are on the collision could be on any radio frequency or no frequency at all. The controller only has one frequency to talk out on, so unless one or the other aircraft happen to be on their frequency, they can do nothing but watch the two targets merge on the screen. They pause a moment, and then almost always the two targets separate and go on their way. The controller has no way of warning the pilots, and they will never know if the pilots saw each other or if it was just luck that prevented a midair collision. The controllers call this the "big sky" theory. This means that the sky is real big, there is lots of room up there, and the chances are good that two airplanes won't hit. If one or the other airplane has an operating Mode C Transponder, then the controllers can know a little bit more about how close the airplanes came, but they can do nothing to prevent a collision. When a controller is working with a pilot on the same frequency, then and only then can a warning be issued, but even then the controller is not required to do so. Traffic alerts come from controllers on a "time-share" basis. That means if they have time they will issue an alert, but if they get busy with something else that has a higher priority, no alert will be given.

You probably have heard about the five-mile ring. The new computer radar displays can place a five-mile radius around a target and no other aircraft are supposed to penetrate that ring, but this five-mile ring only is applied to two IFR airplanes all talking to a controller. There is no five-mile ring around VFR-to-VFR aircraft, or IFR-to-VFR aircraft. The only thing that keeps VFR-to-VFR aircraft and IFR-to-VFR aircraft from hitting each other is the pilot's eyes.

The challenge now becomes seeing the traffic and avoiding situations where traffic cannot be seen. A pilot's eyesight is extremely valuable. But good eyesight alone will not guarantee vigilance. A pilot may have 20/20 uncorrected vision but not know how to use his or her eyes properly.

During flight, the pilot must divide attention between what is outside and what is inside. Use a personal "time-share" arrangement with your eyes. Do not spend long periods of time on any one item. Follow a pattern similar to this:

1. Look outside to the left and rear.
2. Look at the flight instruments.
3. Look outside down the left wing.
4. Look at the flight instruments, or make a notation on the nav log.
5. Look outside over the instrument panel out the front.

6. Look at the engine instrument, or check a chart position.
7. Look outside down the right wing.
8. Look at the flight instrument, or change a frequency.
9. Look outside to the right rear.
10. Start over again with number one and repeat.

The more proficient the pilot is in working the radios, using charts, using the Navigation Record, and identifying problems on the instrument panel, the more time is left over for looking outside. When the pilot looks outside, he or she should not "sweep" the sky, but rather look in one direction long enough for the eyes to focus. The eyes do not focus while they are in motion. If pilots look up momentarily from the cockpit and quickly pass their eyes from left to right in the name of collision avoidance, nothing has been accomplished. Even 20/20 eyesight will not help if the eyes are not allowed to stop and focus. When the eyes are directed outside, they will focus on the first thing in the field of vision. If the first thing is a bug on the windshield, the eyes will focus on the bug and temporarily miss what is on the other side of the bug, namely another airplane.

Imagine for a moment two airplanes that are approaching head-on. The regulations say, *"When aircraft are approaching head-on or nearly so, each pilot shall alter course to the right."* This works great if the pilots see each other. But what if they do not see each other in time to alter course?

The two airplanes are three miles apart. First think of the two airplanes as single-engine trainers, each with a cruise speed near 100 knots. As these two airplanes approach head-on their closure rate will be 200 knots. At 200 knots how long will it take the airplanes to close from three miles to impact? Collision will occur in 45 seconds. This means that if pilots are not looking for traffic at least every 45 seconds, they are flying blind.

What if the two airplanes were a Piper with a speed of 100 knots and a lear jet with a speed of 250 knots. They would have a closure rate of 350 knots. If the pilots in this situation were flying with only three miles of separation, the time to recognize the problem and take evasive action would be just 26 seconds. If either pilot were scanning as quickly as once around every 15 seconds, then the time interval for evasive maneuvers could only be approximately 11 seconds. If the pilots were scanning only as often as 30 seconds, they could collide without ever seeing each other. Remember, the witnesses to several of the example midair collisions said they saw no evasive action taken. Several midair collision survivors said they did not see the other airplane until it was too late.

Very few midair collisions are head-on. Most happen when an airplane overtakes another from behind or from the side. This means that we must look in all directions constantly.

Sterile Cockpit

Fly with a "sterile cockpit" around airports just like air carrier flight crews do. Below 10,000 feet they limit all flight deck conversation to the problems at hand: a safe approach and landing. In the same way general aviation pilots should stop carrying on normal conversation and start talking only about active runways, pattern entries, and other traffic from 10 miles into the airport.

In small, general aviation airplanes there is no separation between the cockpit and passenger compartment, so everyone, pilots and non-pilots, must follow sterile cockpit rules and all eyes must be outside the airplane. I have seen passengers climb into the back of a small four-seat airplane and pull out a newspaper for the trip. You should let them know that they are not flying coach, and they will be asked to be a participant and be a lookout throughout the flight. Every person in a general aviation airplane must become an observer member of the crew. I knew a pilot once that began every preflight briefing by pulling out a dollar bill and sticking it in between the radios. He said that during the flight if anybody points out another airplane before he saw it, they would win the dollar. That was money well spent.

I drive to work each day on a highway that is known for its bad accidents. It's a narrow, two-lane, winding, state highway. I know before I even leave the house that I have a greater chance of having an accident on that stretch of road than any other place. Knowing this I really try to concentrate more when the danger is greater. As pilots we know the greatest danger of a midair collision exists around the airport, in the traffic pattern, and especially on final approach. Approaching an airport is just like driving down that dangerous highway. We know the danger is increased so we must increase our concentration, our lookout, and our radio work. You should never let your guard down when it comes to looking for traffic, but when you know the danger is present, you must raise your level of vigilance.

Killing Zone Survivor Story

NASA Number: 409892

I was returning VFR to my home base. I was receiving flight following from an approach control. After being handed off, I called 5.8 nautical miles southwest of the field. I asked for an advisory from unicom (123.0), and a pilot in the traffic pattern called the pattern for runway 15. I announced I would enter downwind for runway 15. I had just departed the same airport on runway 33 about 45 minutes prior. I proceeded to enter downwind and base for runway 33. I had been busy messing with the autopilot just prior to the handoff from approach, and had a mindset for landing on runway 33. As I entered final for runway 33 a pilot departing runway 15 flew above me and banked to the right. He radioed to tell me that I was on runway 33 not 15. Even though I said "runway 15" I actually flew a perfect pattern for runway 33.

CHAPTER NINE

Fuel Management

FUEL-RELATED ACCIDENTS divide up into three groups.

1. Accidents that involve fuel contamination.
2. Fuel exhaustion.
3. Fuel starvation.

The contamination of fuel is mostly from water. In 1998, the Air Safety foundation reported that there were a total of 12 fuel contamination accidents, with two of these resulting in a fatal accident. The following is the report from one of these accidents—one of the 10 nonfatal.

NTSB Number: ATL99LA034. Shelbyville, Tennessee

On December 14, 1998 at 1000 Central Standard Time, a Cessna 152, collided with the ground, according to the student pilot's flight instructor, during an attempted forced landing ten miles south of Shelbyville, Tennessee. The student pilot had filed a flight plan for a solo cross-country flight. An examination of the airplane at the accident site disclosed that the airframe had substantial structural damage. The student pilot was not injured.

The student reported that, approximately forty-five minutes into the flight, the engine rpm dropped. The student pulled the throttle lever to the idle position, and applied carburetor heat. The engine rpm continued to drop, and the engine subsequently quit. Attempts by the student to restore full engine power failed. The student selected an open field for an emergency landing. During the landing attempt, the nose wheel assembly sustained structural damage.

Examination of the airplane disclosed that the nose gear assembly had sheared off and that the engine firewall had sustained impact damage. The examination also revealed that a fuel line in the vicinity of the chin cowling had also ruptured and an unknown quantity of fuel was spilled. Approximately a quart of liquid was recovered from the fuel system; 90% of the liquid was clear water. Water was also discovered in the carburetor bowl, and the filter assembly.

Reportedly the airplane was refueled about two days before the accident, and subsequent to the refueling the area experienced severe thunderstorm activity. Security of the fuel caps after the refueling was not confirmed. The pilot did not report checking and securing the fuel caps during the preflight inspection.

Probable Cause:

Water contamination of the fuel supply, resulting from the pilot's inadequate preflight inspection.

Figures 9.1a and 9.1b are photographs from this accident. The pilot attempted to make a forced landing in the field, but as you can see in Figure 9.1a the field has an incline from left to right. The pilot landed as if the field were flat, and from the air it surely looked flat, but the incline caused the nose wheel to touch down first. It touched down with enough force to break and bend back the nose gear. The airplane then slid like a wheelbarrow with no front wheel uphill until it came to rest. The pilot was unhurt. The accident narrative mentioned the unconfirmed security of the fuel caps, and the probable cause places the responsibility for the accident on the "pilot's inadequate preflight inspection." But you can see in Figure 9.1b that the fuel caps are secure and there was over 15 gallons of fuel in the airplane. The report mentions "severe thunderstorm activity" in the local area, but did not men-

Fig. 9.1a The pilot landed and was uninjured. The field selected sloped upward. The pilot, unaccustomed to landing on an incline, did not flare sufficiently and broke off the nose gear when touching down.

tion the fact that the night before the flight the temperature had dropped well below freezing.

On the morning of this flight the student pilot and his flight instructor both walked out to the airplane. When they arrived both fuel caps were off, but the fuel truck had just pulled away. They assumed that the line service had just finished filling the fuel tanks, and that the caps had been left off as a courtesy to the pilot. In this way, the pilot could confirm that the tanks were full and then personally secure the caps. This assumption turned out to be the true cause of the accident.

It was later discovered that the fuel truck in fact had not filled the tanks that morning. The truck was driven to the airplane but the lineman found the tanks already full and drove away. The student pilot, instructor, and lineman did not speak to each other, as the lineman continued on to other airplanes. The lineman saw that the airplane was full but did not replace the caps, thinking the pilots were on their way and they would want to check the caps themselves. This meant that the caps had been off longer than the instructor believed. Both instructor and student pilot thought the caps had only been off for the time it took to

Fig. 9.1b The fuel caps were fastened properly for the flight.

refuel the airplane that morning, when in fact the caps had been off prior to the lineman's arrival at the airplane. How long had the caps been off and why were they off in the first place?

Further investigation revealed that the airplane had been last filled with fuel two days before—prior to the heavy rain. Could the caps have been left off for two days allowing rain into the tanks? Yes. The lineman who had worked two days earlier reported that he had filled the tanks, and because the pilots were present at the fueling, he had deliberately left the caps off. Then the pilots decided against the flight. They gathered up all their gear and went back inside—not knowing that the lineman had left the caps off for their inspection. So there the airplane sat for two days in the rain.

On the morning of the accident the student pilot and instructor were together when the two main tanks and the belly sump were drained. The temperature during the preflight inspection was still below freezing, but both later reported that uncontaminated fuel came from all three sumps and all seemed normal. I always thought that since water is heavier than fuel that any and all water would collect at the lowest point in the tank where we can drain it out. This is true, but the inside of the tank is not completely smooth. There are places where water can hide. Was it possible that the rain water that had entered the tank over a two-day period had accumulated and then froze in an area away from the drain? When the fuel sample was taken, uncontaminated fuel came out, but ice was nevertheless in the tank.

The accident happened at around 10 o'clock in the morning. From Figure 9.1a you can see that there were no clouds in the sky and the sun was warming the earth. Temperature at the time of takeoff was rising higher than the freezing mark. Most fuel contamination accidents happen during the takeoff run or soon after takeoff. This time interval from start-up allows any water that might be in the system to work its way to the carburetor. But the first sign of an engine problem in this situation did not take place until 45 minutes into a cross-country flight. Did the rising temperatures melt the ice in the tanks, and in effect contaminate the fuel en route? The report says, *"Approximately a quart of liquid was recovered from the fuel system; 90% of the liquid was clear water. Water was also discovered in the carburetor bowl, and the filter assembly."* If the fuel sumps were drained before departure and no water was found, where did the water come from 45 minutes later?

Be very careful when temperatures are rising from below freezing overnight. It is a good idea to walk to the end of the wing and gently rock the wing up and down. This might help move water to the lowest point in the tank and fuel system, where it will come out in the drain. And always replace fuel caps. The lineman thought he was doing the pilots a favor by leaving the caps off so the pilot could verify fuel quantity for themselves, but in this case a chain of events started two days before the accident. The caps left off, the rain, the subfreezing temperatures the night before the flight—all of these links in the chain came together to cause an accident.

The next example involved an airplane that did lose power on the takeoff run. Later, water was again found in the fuel system.

NTSB Number: MIA91FA210. Miami, Florida

Witnesses reported hearing the engine run rough while on a takeoff roll but the flight continued to takeoff. Shortly after becoming airborne the pilot radioed the tower that he was having engine problems and was executing an off airport forced landing. The airplane was observed to turn to a suitable landing area and initiated a descent. At about 100 AGL the airplane banked to the left and turned into a residential area. The airplane crashed into a residence and burst into flames. Examination of the engine revealed evidence of water ingestion. The airplane had not been flown for a week and was stored outside. The rainfall the previous week had been heavy.

Probable Cause:

The failure of the pilot to adequately drain the fuel system of water. A contributing factor was the pilot's decision to continue takeoff with a known deficiency. The pilot's selection of a forced landing area was poor in that a better landing area was available.

There have been several accidents where a rough-running engine started on the takeoff run, but the pilot kept going. One pilot said later that he could not hear the engine roughness due to the quality of his new headset. We have done such a good job with headsets in reducing noise that noises that we really need to hear have been covered up. I am all for the use of headsets, but this may become a bigger problem as headsets get better.

This accident had some similarities to the first accident example. Both airplanes had been tied down outside. Neither airplane had been flown for a few days, and rain had been in both areas since the airplane's last flight. The probable cause also mentioned "failure of the pilot to adequately drain the fuel system of water." There was one difference; the second example happened in Miami in August so there was no threat of ice.

Remember the basics of fuel testing and test each and every time. Drain a sample from every tank and every sump. When the sample is drained, don't just throw it away. First hold the clear sample cup up against a white surface. I hold the cup up against the white part of the airplane. Doing this will make it easy to see if the color of the fuel is correct. Today just about all you can get is Aviation 100LL, which is tinted blue. Next, smell the sample and be certain it is gasoline, not kerosene. It would be rare for a lineman to mistake a Cessna 152 for a turboprop, but stranger things have happened and the wrong fuel can get into the wrong airplane. The word "turbo" has been written on airplanes to mean that the engine is a "turbocharged" piston engine. But "turbo" has been misinterpreted to mean "turboprop," which is a turbine engine. I routinely fly a 10-passenger airplane that has piston engines. Twice before, both times at large airports, a lineman has driven a JetA truck up to that airplane thinking it was a turboprop. It usually pays to be present when your airplane is being serviced.

When you take the sample of fuel, make sure it is indeed fuel before you discard it. It may be an entire cup of water. When a small amount of water is captured in the drain cup, it looks like bubbles on the bottom of the cup and this is easy to see. But a full cup with no bubbles might look like clean fuel, when in fact it is clean water. Always check for color and smell.

Even though I check items in the order of the walk-around inspection, I always go back and double check that baggage doors are locked, oil dip sticks are back in place, and fuel caps are on and secure. These are the last things I do before getting in the airplane.

NTSB Number: BFO89FA062. Cluster Springs, Virginia

The student pilot was en route when she radioed the engine was losing power. According to another pilot, the student sounded panic-stricken during the transmission. The aircraft crashed into a small pond, surrounded by an open field. Damage to the aircraft, indicated the aircraft struck the pond in a steep nose down attitude. Examination of the aircraft did not reveal evidence of mechanical malfunction. Inspection of the left wing surface behind the fuel tank filler neck, indicated the fuel cap was unfastened and laying on the wing inflight. A flight test in a similar aircraft with an unfastened left fuel cap revealed the cap moved aft and was held by its chain. During the test, fuel was siphoned from the left fuel tank. Fuel performance calculations indicated the aircraft would have used up to 13 gallons during the flight. It had a 30 gallon fuel capacity, 15 gallons per tank.

Probable Cause:

The pilot's failure to perform an adequate preflight inspection that resulted in an unlatched fuel filler cap. Contributing to the accident was the pilot's failure to monitor the fuel supply during the flight.

This tragedy took place in a Beech 77 Skipper. The student pilot with 103 hours was killed. The wing produces lift by creating low pressure on the upper camber of the wing. But the fuel filler neck is also on the upper camber of the wing. If the fuel caps are left off, the same low pressure that produces lift for flight will suck the fuel out of the tank like a straw.

NTSB Number: CHI96FA073. Hallam, Nebraska

The airplane departed Dallas, Texas at 1544 Central Standard Time with a gross weight of 2,382.3 pounds. The maximum certified gross weight was 2,081 pounds. About 4 hours and 15 minutes after takeoff, the pilot transmitted on a unicom frequency that the airplane was low on fuel. A short time later, he reported that the airplane was out of fuel and he intended to perform an off-airport landing. During a night forced landing, the airplane contacted powerlines, then impacted the terrain. A total of 30 ounces of fuel was drained from the airplane fuel system after the accident.

Probable Cause:

Improper planning/decision by the pilot, which resulted in fuel exhaustion, due to an inadequate supply of fuel. Darkness was a related factor.

This accident took place in a Piper Cherokee 140. The private pilot with 139 hours suffered fatal injuries. This was classified as a fuel exhaustion accident. Fuel exhaustion takes place when no more fuel is left in the airplane. This airplane was found with 30 *ounces* of fuel remaining after a 4-hour and 15-minute flight. Did the pilot lose track of time? Did the pilot think that there was extra fuel since the takeoff was made overweight? Did the fuel gauges make it look like there was more fuel than there actually was? The only person who could answer these questions did not survive. But the fuel gauge question needs to be addressed. I always heard that fuel gauges were not required to be accurate unless they were empty, but I went most of my flying career thinking that was just hangar talk. It turns out that it is true, but the regulation that allows it is very hard to find. The rule is from the old Civil Aviation Regulations (CARs) and has never been superseded. Specifically, it is CAR 3.672 but probably could only be found on microfilm at a well-stocked library. The rule does say that the only time a fuel gauge must tell the truth is when it is empty. With any quantity of fuel in the tanks the gauge reading is just a ballpark reading. Therefore, you should never trust the fuel tank gauges. Instead use time of fuel burn. The pilot of the previous accident example should have landed before 4 hours and 15 minutes had passed, even if the gauges indicated some fuel remaining.

Don't ever take off without knowing just how much fuel is in the airplane. I prefer to "top off" the tanks because that way I know exactly how much I have. I do not rely on any "stick your finger in the tank method." I have heard pilots say, "It's OK. I can touch the fuel." No, that's not OK. If for weight reasons, you elect to take off without full tanks, then stick the tank with a calibrated tank indicator. You can buy or make these.

A flight instructor was assigned by his chief pilot to give a biannual flight review to a pilot one day. The instructor and BFR pilot had never met before this day. Before the flight the pilot went out to begin the preflight inspection. When the instructor arrived on the scene the pilot was up on the wing checking the fuel visually. The instructor said, "How's the fuel?" and the pilot responded, "It's OK." One hour and 20 minutes into the flight the airplane ran out of fuel. What did the pilot mean when he said the fuel was "OK?" Did he think it was OK because he could reach down into the tank and touch the fuel, or by some sight estimate? What about the instructor? Shouldn't he have asked, "How many gallons do we have?" instead of "How's the fuel?" Neither pilot was hurt during the forced landing, and they almost glided back to the airport. But they did not make it; they hit the airport security fence and the airplane was destroyed. The instructor's resume was also destroyed. He had an upcoming interview with American Airlines that was canceled when the accident appeared in his FAA file. Not being sure about the fuel quantity that day was literally a million-dollar mistake.

NTSB Number: FTW89FA155. Pauls Valley, Oklahoma

The aircraft had been refueled at a nearby airport before this pilot started his flight. He took off and flew the aircraft from Oklahoma City to Tishomingo, Oklahoma, with the fuel selector positioned to the right tank (the tank that was last used by the previous pilot). On the return flight to Oklahoma City at night, the pilot's father noticed that right fuel gauge indicated empty while the left indicated full. The pilot could not find the fuel selector valve and requested help from the Air Route Traffic Control Center. Before they could help him, the engine lost power. There happened to be a pilot at the ARTCC, who was familiar with the PA-28. By the time he was found and the information was passed to the pilot, the aircraft had nearly descended to the ground. As the pilot was reaching for the fuel selector, the aircraft hit trees and crashed. Later, the pilot said he had never operated the landing gear or fuel selector during his checkout in the aircraft. He thought the fuel selector was on the floor, between the seats, but it was located on the left wall of the cockpit, near the pilot's left knee. The fuel selector was found positioned to the empty right tank.

Probable Cause:

Improper planning/decision by the pilot, his lack of understanding of the procedures for operating the aircraft, his improper use of the fuel selector in managing his fuel supply, fuel starvation, and the pilot's failure to know and follow emergency procedures when the engine lost power. Contributing factors were: inadequate transition training provided by the instructor pilot, the pilot's lack of familiarity with the aircraft, the dark night, and the trees.

Fuel starvation is the classification when the engine stops, but there was fuel remaining somewhere in the airplane. Sometimes a fuel system component can fail, making it impossible to deliver fuel from an outlying tank to an engine. Sometimes, like in this accident example, it is just a pilot who does not know how to switch tanks. This accident took place in a Piper Cherokee 180 with three aboard. The pilot, who had 83 flight hours, and another passenger survived with serious injuries. But the pilot's father, who first saw the lopsided fuel gauge readings, was killed.

Across all accident categories there has been a common theme. You must know the aircraft you fly in order to fly it safely. Knowing the airplane should not stop with an insurance checkout or a skimming of the pilot's operating handbook. All pilots should know how things work in the airplanes they fly. At a minimum, knowledge of the ignition, electrical, hydraulic, flight control, and fuel systems are imperative.

NTSB Number: LAX91FA106. Fredonia, Arizona

The 22-year old pilot and his three passengers departed on a three-day excursion of Bryce and Grand Canyons. The pilot had logged about 72 hours of flight time. The rented aircraft departed at night 68 pounds over gross weight and one inch aft of the most rearward allowable center of gravity limit. After reaching their first destination and spending the day sightseeing, they departed again at night for the Grand Canyon. The weather was marginal with snow showers. The planned 56 minute leg was tracked on radar for a total of 1 hour and 39 minutes, a meandering and undulating track up to 13,400 feet MSL was recorded. The aircraft departed with 1 hour and 40 minutes of fuel in each tank. At the accident site the right tank was full and the left tank was empty. The fuel selector was on the left tank. The aircraft impacted mountainous terrain at about 7,000 feet MSL in a flat spin.

Probable Cause:

The pilot's inattention to the airspeed and subsequent loss of aircraft control after attempting VFR flight into adverse weather conditions. Contributing to the accident was the pilot's failure to recognize his limitations and the limitations of the aircraft.

Four people died in this accident that was so avoidable. Don't hike all day and then go flying. Don't even work all day and then go flying into the night. The pilot got lost, which took up more time and burned more fuel. He either forgot to switch fuel tanks or did not know to switch tanks. When the engine stopped the pilot became distracted, attempting to find the fuel selector. During the struggle, airspeed was lost and the airplane was allowed to stall and then spin.

NTSB Number: MIA92FA173. Picayune, Mississippi

The private pilot and one passenger departed on a two-hour flight with about 20 gallons of fuel in each tank. The airplane was not found until two days later, burned in a wooded area. The fuel consumption for two hours would be about 19 gallons. The fuel selector was found on the left tank and the left wing exhibited less fire damage than the right wing. No evidence of powerplant or airframe preexisting failures were found.

Probable Cause:

The failure of the pilot in command to place the fuel selector on the fullest tank, resulting in an engine failure due to fuel starvation over unsuitable terrain.

This Beech A23-19 had less fire damage on the left side because there was no fuel in the left tank. It is terrible when an engine stops and then an airplane crashes when there is still a full tank of fuel on board. In this case the private pilot with 75 hours was killed, along with one passenger.

Very often, accident causes overlap. There are many examples of fuel mismanagement coupled with stall/spin accidents. When fuel supply is interrupted to the engine, the engine quits or starts acting like it is going to quit, and this severely distracts the pilot. As the pilot frantically attempts to come to grips with the situation, the pilot's number-one job is ignored: flying the airplane. The pilots turn their attention to the crisis and away from the airplane. If flying the airplane is completely neglected, soon the fuel will be the least of the pilot's problems. Having an engine quit due to a lack of fuel is bad. Being out of control is worse.

NTSB Number: MIA90FA106. Marco Island, Florida

The pilot departed Naples Municipal Airport, in Naples, Florida, VFR on a personal flight at 11:30 hours. The airplane was observed conducting touch-and-go landings at the Marco Island Municipal Airport in Marco Island, Florida between 1300 and 1400 hours and then departed at an unknown time. The airplane wreckage was initially spotted at 1700 hours in a bay, but was thought to be a sunken boat. The Civil Air Patrol spotted the airplane at 1925 hours and notified the authorities, who responded to the crash site. Aircraft records and Hobbs meter in the airplane, indicate it had been operated 4 hours and 36 minutes. Aircraft performance charts indicate fuel exhaustion at 4 hours and 24 minutes. There was no evidence of fuel at the crash site or purchase of fuel by the pilot. Physical evidence indicates inflight loss of control. The airplane stalled and spun to the left and impacted the water. There was no evidence of any powerplant, airframe, or flight control failure.

Probable Cause:

Fuel exhaustion due to improper planning and decision by the pilot, and an inadvertent stall/spin while maneuvering for a forced landing.

In the event of an engine failure, use this phrase to attack the problem: Fly-Find-Fix. First and always the pilot must fly the airplane. In this case it means the pilot must maintain airspeed and avoid a stall. If the engine quits, the airplane becomes a glider. The failure of the engine does not cancel aerodynamics. The airplane will still create lift if the pilot flies the airplane. Next, find a place to land. If you are flying over rough or mountainous terrain, there may not be any good choices. This should be a factor when planning a route of flight. If there is no place to land, should an engine fail, maybe you should go a different way. If there is a place to land, attempt to maneuver into a "normal" traffic pattern. You will have better judgment about height if you can make a downwind, base and final approach to the field. You already have seen traffic pattern approaches countless times, so put that judgment to work for you in an emergency. Remember to land into the wind if possible. Last, attempt to fix the airplane. This means check carburetor heat, switch fuel tanks, check that the magnetos are on, and/or adjust the mixture setting. Each airplane has a troubleshooting checklist.

Solving the problem with Fly-Find-Fix will give you the best opportunity for success. If these are taken out of order, if fix is attempted first without getting the airplane under control, then the pilot could lose control altogether.

NTSB Number: MIA93FA065. Aguadilla, Puerto Rico

The day before the accident flight, 12 gallons of fuel were added to the fuel tanks and the airplane flew three times. The accident flight departed with an unknown quantity of fuel and while maneuvering, the engine quit due to fuel exhaustion. During the emergency descent, one of the two occupants applied full-up elevator control at an airspeed above maneuvering speed (Va) which caused both forward wing spars to fail in the positive direction. The airplane descended near vertical and impacted the ground in a nose low attitude. Exam of the aircraft at the accident site revealed no evidence of flight control preimpact failure or malfunction. Exam of the fuel system revealed no evidence of fuel. Metallurgical examination of the fracture surfaces of the wing spars revealed no evidence of preexisting cracks or corrosion. The airplane had been operated for approximately 30 minutes since departure. The student pilot in the left seat was the only occupant who was authorized to fly the airplane. No determination could be made as to why the private pilot rated passenger was in the airplane.

Probable Cause:

Excessive elevator control input by an unknown occupant at an airspeed above maneuvering speed that caused both forward wing spars to fail. Contributing to the accident was fuel exhaustion due to poor preflight planning/decision by the pilot in command and his total lack of experience.

If the engine quits, the airplane must still be flown properly. For some unknown reason the pilot of this Ercoupe F-1 tore off the wings with their violent movement of the control wheel. There was still some question as to which pilot pulled the wheel. The student pilot, with 15 hours, was in the left seat. If he was acting as pilot in command, then he was illegally carrying a passenger. The private pilot in the right seat could have been acting as pilot in command and the student pilot was his passenger, but ordinarily the PIC sits on the left side. We will not know for sure; both pilots were killed.

Several years ago a pilot ran out of fuel on the way to a NASCAR race. He had three friends with him in the airplane. When the engine quit it was just getting dark, but the pilot managed to land in a field with minor injuries. The local news media arrived on the scene and interviewed the pilot. The next day on the evening newscast the pilot, standing beside the airplane wreckage, said, "I don't know what the problem was. Me and my buddies fly to the race every year, and we always make it."

Obviously this pilot did not calculate any fuel consumption versus his ground speed. Just because you make it from point A to point B on one day does not mean you will make it another day if the wind is different. This is a fact that pilots just should know. It seemed incredible that a pilot would have made such a statement. Later that week I was teaching an instrument pilot ground school at night. We happened to be talking about IFR flight planning that week, and how fuel calculations are important for VFR but vital for IFR. I used the example of the pilot who ran out of fuel on the way to race as an object lesson. I probably said things like, "How could this pilot have made such an error? He really must not have been thinking at all." I went on with the class; eventually I wrapped up and the students started filing out. The last guy to leave the class was a man who had been sitting in the back. Instead of leaving he came to the front of the room. That's when I noticed he had a big black eye. You guessed it. It was the pilot who ran out of fuel. "Hey man, that was us," he said, "but you don't have to go around talking about it."

Wind has a direct influence of fuel consumption. It is clear that if you have a headwind, the airplane will travel across the ground slower. This means it takes longer to get where you're going. This means you will be in the air with the engine running longer, and this means it will burn more fuel. Pilots should always time cross-country legs. Before takeoff, calculate your ground speed under the prevailing wind. Then measure the distance between checkpoints all along the way. With the ground speed and distance known, you can work the problem to discover how long it should take to get there. When you arrive over a checkpoint, compare how long it has actually taken against how long you thought it should take. Most of the time your calculations will be very close to reality, but sometimes it will be different. If you got there faster than expected—great! Maybe you have a tailwind or less of a headwind than predicted. But then there will be plenty of flights where the checkpoints are consistently late. Late to a checkpoint means you were going slower than expected and you burned more fuel to get there. If this continues throughout the flight, you will need more fuel than you had planned. When this situation occurs, pilots must be aware and make a fuel stop. Having a different-than-planned ground speed is not all that unusual. Remember the winds that we use to make these ground speed calculations comes from the winds and temperature aloft forecast. This forecast is usually very accurate, but it is a *forecast* of winds, not actual winds. Forecasts can be wrong. You, while flying in the actual wind, will have better information about the wind than any forecast.

If you do not ever time your legs, you cannot know if you are ahead of schedule and therefore saving fuel, or behind schedule and running out of gas. Make at least two time measurements for each leg.

There is also a peculiar wind effect that most pilots don't realize. You may have heard the saying, "Every wind is a headwind." How can every wind be a headwind? There are in fact times (rare though they may be) when the wind is at our backs. The effects refer to an out and back trip where a pilot flies out with a headwind, and then back again against what then is a tailwind. Doesn't the extra time it takes to fly against the wind get made up with the return trip's tailwind? Should the extra cost of fuel against the wind get paid back with a tailwind? No, it doesn't. Any wind acts like a headwind, because any wind, when you out and back, will cost you more time and fuel than if the wind were calm.

Figure 9.2 depicts a flight from Airport A to Airport B when there is no wind. To make the numbers easy, let's say that it is 100 nautical miles between Airports A and B and that our airplane will fly with a true airspeed (the through-the-air speed) of 100 knots. This means that it would take us exactly one hour over and one hour back, a total travel time of two hours. Now Figure 9.3 illustrates the same trip, but this time there is a 25-knot wind. The wind is a headwind on the way outbound; this means that our ground speed into the teeth of the wind is reduced to 75 knots (100 knots true airspeed minus the 25-knot headwind). Traveling across the ground at 75 knots, it will take us 1 hour and 20 minutes to fly 100 miles to Airport B. But the return trip will have a great tailwind. Will the extra speed coming back make up for the slower speed going out? Our ground speed coming back will be 125 knots (100 knots true airspeed plus 25 knots tailwind). At a speed of 125 knots, it will take 48 minutes to fly 100 nautical miles. Remember, it took exactly 2 hours with no wind. But with a wind it took 1 hour and 20 minutes outbound, and 48 minutes inbound. With wind the trip took a total of 2 hours and 8 minutes. The tailwind did not make up for the headwind.

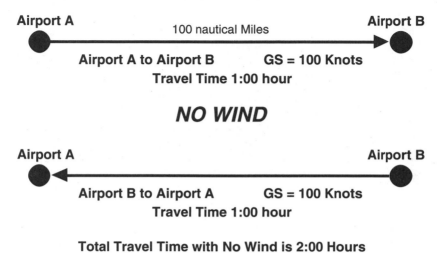

True Airspeed 100 Knots

Airport A   100 nautical Miles   **Airport B**

Airport A to Airport B   GS = 100 Knots
Travel Time 1:00 hour

NO WIND

Airport A   **Airport B**

Airport B to Airport A   GS = 100 Knots
Travel Time 1:00 hour

Total Travel Time with No Wind is 2:00 Hours

Fig. 9.2

It takes more fuel when there is a wind present. On the surface this does not make sense. Why aren't the headwind and tailwind a trade-off? Because when the headwind is present and the ground speed is slower, the wind has more time to act on the airplane. The return-trip tailwind, because it takes less time, has less opportunity to yield an advantage. So any wind, when it comes to fuel calculations, must be considered an increase in fuel consumption. Several pilots who did not understand this fact have run out of fuel and could not understand why.

If you ever are concerned about the amount of fuel in the airplane, you should plan a precautionary landing. That does not necessarily mean an off-airport landing, but a landing as soon as practicable. Air traffic controllers could help you find the nearest airport if they can get your radar identified. When you talk about fuel on the radio, it is important to use the correct phraseology. You should say, "I have a minimum fuel alert" whenever you are concerned enough to make a precautionary landing. A minimum fuel alert is not an emergency declaration, but it will get you quick handling.

Don't take any chances when it comes to fuel. In small airplanes you should take off with full tanks everytime, unless you reduce the fuel amount as a weight consideration. If you do that, you should calculate the weight and balance for the airplane and know how much fuel must be

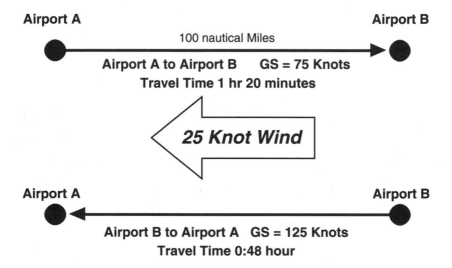

True Airspeed 100 Knots

Airport A   Airport B
100 nautical Miles
Airport A to Airport B   GS = 75 Knots
Travel Time 1 hr 20 minutes

25 Knot Wind

Airport A   Airport B

Airport B to Airport A   GS = 125 Knots
Travel Time 0:48 hour

Total Travel Time with a 25 Kt Wind is 2 Hrs 8 Minutes
Fig. 9.3

reduced. As the weight the airplane carries goes up, the airplane's range goes down. You might be able to make a trip nonstop with full fuel tanks when it's just you and a friend, but you may have to make a fuel stop if you take three friends and camping gear. In order to take off at the proper weight, you have to take out a pound of fuel for every pound of people and gear you add. With every pound of fuel that is left behind, the shorter the leg must be. Do the math each time and get it right. If you have not done a fuel-for-weight problem lately, or if you are planning for an airplane that you are not as familiar with, then get with a flight instructor. Running out of fuel is inexcusable. A little fuel planning and adequate airplane familiarization would eliminate fuel mismanagement accidents.

Killing Zone Survival Stories

NASA Number 409296

A weather briefing from Altoona flight service indicated marginal
VFR during the first part of my VFR flight with improving condi-

tions in the last $^1/_2$ of the trip (after Zanesville, Ohio). Current weather showed stations better than MVFR with possible 2 to 7 miles in fog and mist on the first part of the flight. The trouble spot extended from Wheeling, West Virginia through Zanesville. I calculated ground speed at 90.3 knots with a true airspeed of 105 knots. Total time en route was estimated at 3 hours and 23 minutes over a distance of 306 miles. Total fuel aboard was estimated at 4 hours and 30 minutes. The problem arose en route when I had to take a more westerly course to remain in VFR weather. I stayed west of Zanesville and continued back to my original course south of Zanesville. I lost about 30 minutes by circumventing the problem area. As I got closer to Lexington, I was uncomfortable with the fuel situation; however, I was within the time frame of my fuel estimate. Fifty miles northeast of Lexington, I decided to refuel at FGX. I circled the field and landed but was informed that no fuel was available (the tanks were being replaced). I received this information while landing and decided not to shut down the engine, but to continue on to Lexington. I calculated total flight time would be 4 hours and 10 minutes. Four hours and 5 minutes into the flight, the engine stopped 6 miles short of Lexington airport. I was in contact with Lexington approach and radioed that I was out of fuel and landing in a pasture field below. When on the ground, I used a cell phone to talk to Lexington airport and arrange for fuel to be transported. We estimated the distance of the field and the takeoff performance of the airplane. I loaded 10 gallons of fuel and made a soft field takeoff with no incident and then landed at Lexington Blue Grass airport. In looking back at the situation, I could have avoided this problem with better fuel management inflight and/or by setting a personnel safety factor for reserve fuel. In the future a one hour reserve plus better fuel management will prevent this from happening again.

NASA Number: 409766

I was near to a landing when the engine stopped. My assessment was a lack of fuel. I was told the aircraft had 4 hours usable and was basing my decision on that. I now believe that time was too generous. I landed the aircraft in a stubble field away from people and property and had declared an emergency. The Hobbs Meter indicated 3.4 hours. Landing was made safely without damage to property or aircraft. There were no injuries. The aircraft was inspected and then flown off the highway nearby. Police were there to assist and hold traffic.

These two pilots made a series of very poor decisions. The second pilot said his fuel calculation was based on "what he was told." Who was the pilot in command on that flight anyway? The first pilot had calculated 4 hours and 10 minutes of fuel, but knowing this, was still in the air and seeming upset at the airplane for running out of fuel just 6 miles short. Despite themselves, these pilots were not hurt, but it did not sound like they learned very much.

CHAPTER TEN

Pilot Health, Alcohol, and Drugs

As PILOTS WE ARE accustomed to medical exams. The student pilot certificate does not come from the FAA, a flight instructor, or a pilot examiner; it comes from a doctor. The medical certificate process is designed to insure that health problems will not directly interfere with flight operations. It is rare when a health problem endangers a flight. Here is one example:

NTSB Number: MKC86LA178. Emporia, Kansas

The aircraft was found stopped in an open field having penetrated the southern boundary of the Emporia Airport property. The aircraft engine was still running when rescuers approached the location. A forced entry was made into the aircraft [Cessna 172] to shut the engine down and rescue the pilot in command. Cardiopulmonary resuscitation was attempted and the pilot was taken to a local hospital. The pilot was pronounced dead on arrival at the hospital. Subsequent pathology indicated that the pilot died of an acute hemorrhage into and rupture of a plaque and floppy mitral valve, normally associated with sudden death. Autopsy disclosed no additional factors contributing to the death. The pilot was being treated for hypertension and had been using 250 mg daily dosage of diural medication as part of his treatment.

Thank goodness that this pilot's sudden death did not take place just a few moments later after takeoff. A sudden pilot incapacitation in flight is rare and causes very few accidents. The accidents that relate to pilot health are more normally related to self-induced problems that happen between medical exams. The regulations do address medical problems that might occur between medical exams and ask that pilots "self-regulate" themselves. You may hold a valid medical certificate, but become ill. You may need to take medication. The regulations ask pilots to ground themselves anytime they cannot safely act as pilot in command. I once broke my arm and I was unable to manipulate the aircraft controls until I got a smaller cast. The regulation is 61.53: Prohibition on operations during medical deficiency. It states that a person, "...shall not act as pilot in command, or in any capacity as a required pilot flight crewmember, while the person: (1) knows or has reason to know of any medical condition that would make the person unable to meet the requirements of a medical certificate necessary for pilot operations; or (2) is taking any medication or receiving medical treatment for a medical condition that results in the person being unable to meet the requirements of a medical certificate necessary for pilot operations." This means that if you experience anything that can cause physical deficiency, then you are supposed to take yourself out of the game. This does not mean that you must give up your medical certificate; it just means that while you have the flu and are taking medicine, you should have the good sense not to fly.

The accidents that pertain to pilot health or readiness to fly break down into fatigue, use of over-the-counter and prescription drugs, use of alcohol, and use of illegal drugs.

Fatigue

The airlines, charter pilots, and flight instructors have duty time requirements. These regulations are designed to limit a pilot's workday so that fatigue does not create a safety problem. In 2000 the FAA changed airline reserve duty time requirements after two tired pilots slid off a runway in Little Rock, Arkansas. A reserve pilot is not scheduled to fly a particular route, but is used as a backup to cover any other pilots who call in sick or are not in the right place at the right time due to cancellations and delays. Before 2000 the reserve pilot had to be within 45 minutes of the flight line for a 24-hour period waiting to go to work if needed. That meant that a pilot could be called in for a flight in the 23rd hour of the day. The flight itself might be an 8-hour work period. If the pilots had known they would be working those eight hours, they could have tried to go to sleep earlier, but they could not have known, and it's not easy to just go to sleep anytime. So the reserve pilots often find themselves staying awake for more than 24 and 30 hours, with the most demanding workload at the end of the period. Approximately 20% of an airline's pilots are on reserve at any time.

The rule change made in 2000 now stops pilots from working beyond 24 hours. If pilots are on reserve and are called in to work, the flight they are taking must end by the end of the 24-hour period. If a flight were to take 6 hours (a flight can be many stops and individual legs), then a reserve pilot would have to be called in before the 17th hour of the 24-hour period. The remaining 7 hours of available duty time that the reserve pilot has to give will be enough to get to the airport and make the flight. But now, after the 17th hour, that reserve pilot could not take that 6-hour flight. This immediately meant that airlines needed more pilots on reserve and this fact perpetuated the unprecedented pilot hiring boom of the late 1990s.

General aviation pilots flying for business or pleasure have no such duty time requirements and it would be very tough to write a rule to cover this situation. The typical fatigue-related general aviation accident involves a pilot who works all day at a nonpilot job, and then goes flying after already completing a full day. This often happens on the last day of a work week and the pilot is desperately trying to get away for the weekend. Their 7:30 to 6:00 job, full of work-related stress, goes on as usual, and when they finally tie up all the loose ends they are taking off at sunset. Recall the accident example from the previous chapter where a pilot and his friends spent that day hiking and climbing, then getting in the airplane after dark. You cannot be at your best when you are 10 or 12 hours into your day when you first get in the airplane.

NTSB Number: SEA94FA129. Rock Springs, Wyoming

Even though he was tired from flying approximately seven hours earlier in the day, the pilot decided to conduct a local sight-seeing flight with his two passengers. When the flight returned to the vicinity of the airport after dark, thunderstorms were building in the area, and there were strong gusty winds. While executing a

steep descending turn from base to final, the pilot allowed the aircraft's descent rate to become too great, and the left wing of the aircraft impacted the terrain.

Probable Cause:

An excessive descent rate in the turn from base to final. Factors include a dark night, and the pilot's fatigue from his flight and ground schedule the day of the accident.

The pilot was flying a Cessna 182B and was a private pilot with 115 total flight hours. There were other factors involved in this accident, but the probable cause specifically mentions the pilot's fatigue due to the pilot's long day. The pilot had flown 7 hours that day before the accident flight. But that does not mean he had only been working 7 hours that day. Flight instructors are limited to 8 hours of instruction within any 24-hour period. As a person who has tried to make a living as a flight instructor, there have been many days when I came up close to that 8-hour limit. I know that in order to turn 8 hours over on the Hobbs meter, you will have to be at the airport for about 12 hours. Remember, flight instructors do not fly 8 continuous hours. Those 8 hours would come from 6 or 7 individual flight lessons. Each lesson will have preflight and postflight discussions, there is fueling, dispatching, and lunch from a candy machine. An 8-hour flight day can easily turn into a 12 to 14-hour workday.

NTSB Number: LAX91FA040. Kingman, Arizona

The 57-year old pilot departed late in the day for a 7 to 8 hour cross-country flight to beat some incoming weather. He had been pheasant hunting for two days prior. The second leg of his trip was a dark night departure for which he had filed a VFR flight plan and indicated 12,500 feet MSL as his cruising altitude. Radar data disclosed that he climbed 13,100 feet at one point; however, his overall route of flight for nearly three hours was one of a meandering and undulating track. The approximate last 10 minutes of flight was a gradual on-track descent to impact at about 5,700 feet, about one hour short of his destination. He carried no supplemental oxygen. Friends indicated he frequently took *"cat naps" on long cross-country trips while letting the auto-pilot fly the aircraft. His autopilot was operational except for the altitude hold.*

Probable Cause:

Proper altitude not maintained, contributing to the accident was: the lack of supplemental oxygen on board the aircraft, pilot fatigue, improper use of the autopilot, and pilot complacency.

A pilot flying drowsy can be as dangerous as a pilot flying drunk. Pilots must use good sense when planning their daily schedules that involve flying. There are no rest time requirements for personnel flying, but use what the pros do as a guide. One of my former students makes a 10-hour flight twice a month to Sao Paulo, Brazil, as a Boeing 767 captain for United Airlines. Since the flight is longer than their allowed duty time, they must take along extra pilots, so that they can trade off along the way. They must take along a pilot that is called the "meal-eater." This pilot usually stays back in the first-class section until the flight is under way for two or three hours. Then the "meal-eater" pilot comes forward to give one of the pilots a duty time break, but before reaching the destination, the "meal-eater" pilot returns to first class for dinner. They don't take off, they don't land, they just fly straight and level and then eat. And duty time applies to flight attendant crews as well. Once on his return trip from Sao Paulo, the flight attendants that were supposed to work the flight had not arrived. The flight attendant crew that had just come from New York would have passed their duty time limit so they could not continue. The flight had pilots ready to go, but they had to off-load the passengers because no flight attendant crew was present with enough duty time to make the trip. Then when all the passengers were off, the airplane left the gate and flew on to New York. The flight had to be made anyway to get the cargo that was in the airplane delivered and to get the airplane back for the next flight. Can you imagine how upset you might have been if you got tossed off the airplane, and then the airplane went on without you? It is all to limit the effect of fatigue on flight safety. If the pros go to those lengths, then as individual pilots we should also be smart and fly only when we are alert, fresh, and rested.

Killing Zone Survivor Story

NASA Number: 409069

"I was at a fly-in. I was departing late Saturday afternoon. I had camped at the fly-in for two days and had slept poorly. I was tired

and had drunk no water that afternoon. I had flown the evening before and had been distracted from changing from my auxiliary fuel tank to a main wing tank in the traffic pattern prior to landing by heavy traffic. I preflighted prior to departure and checked the fuel in the wing but could not visually check the auxiliary tank. I was #5 in a flight of 5. I had trouble with my battery on start and this delayed my taxi. I reached the run-up area and did my run-up without my checklist and missed selecting fuel to the wing tank (I was rushed because the last plane in the flight was waiting for me). I usually do a configuration check prior to takeoff: trim, magnetos, carb heat, flaps, and fuel. I got to flaps and stopped, the runway director was directing me at that time. I took off on the auxiliary tank. I ran out of fuel at 300 feet after I retracted the gear and flaps. I switched to a wing tank and started pumping the wobble fuel pump but the engine didn't pick up. I selected the landing gear down but they didn't get down prior to landing on the runway. Touched down with gear in transit. The engine stopped and the props curled, but no injury. Lesson: (1) never run any fuel tank below $1/4$. (2) at unusual events double-check yourself. (3) Take all the time you need, use your checklist and don't omit your usual procedure. (4) Ensure that you are rested and physically fit."

The survivor story was taken from the Aviation Safety Reporting System. This pilot was tired, rushed, and neglected the checklist. He did not mention it, but he raised the gear while there was still runway ahead. Had he waited until there was no more available runway to raise the landing gear, he could have made a quasinormal landing right back down without any aircraft damage.

Stress

A pilot must be able to compartmentalize. This means we must leave earth-bound problems on earth. We cannot fly safely if our thoughts are diverted away from flying and directed to bills that need to be paid or arguments with family members. Don't use an airplane to "get away and think."

NTSB Number: NYC89FA121. Augusta, Maine

Shortly after takeoff, the aircraft crashed into a dirt bluff, 197 feet short of runway 8 and 22 feet below the level of the runway. Witnesses who observed the crash said the aircraft did not deviate or rock wings prior to impact. There was no evidence of preexisting failure or malfunction with either the engine or the airframe. The propeller had deep leading edge marks and chordwise striations. There was no evidence of seat belt usage by the pilot, who was thrown clear of the wreckage. The pilot was an investment banker, who had just been placed on leave with pay by his employer, while his financial transactions were being checked.

Probable Cause:

Failure of the pilot to maintain the proper attitude during the approach. A factor related to the accident was: the pilot's psychological condition.

This was a private pilot with 164 flight hours in a Beech B19. Nobody knows for certain if the pilot was thinking about his financial investigation when he should have been thinking about landing, but it is rare when a probable cause cites a pilot's psychological state. The flight took place in daylight with high clouds and visibility 15 miles.

Over-the-Counter-Drugs

Today there are hundreds of remedies and quick-fix drugs that can be purchased anytime. Something as seemingly harmless as aspirin can have negative effects when mixed with flying. Remember that flying takes the body to a higher altitude where there is less oxygen to begin with. If you take an over-the-counter medication, there may be no adverse effect on the ground, but when combined with the oxygen reduction of altitude the drug can react differently on your system.

NTSB Number: CHI95FA114. Rush City, Minnesota

A witness reported that she observed the airplane as it entered the traffic pattern. She described the airplane bouncing and rolling, and it "wobbled" from side to side, then descended abruptly toward the ground. The airport manager estimated the winds were "from the southeast and gusting to about thirty knots." Examination of the wreckage revealed no evidence of pre-impact malfunction. The pilot had been issued a private pilot certificate 7 months prior to the accident. At the time of application the pilot listed 64.7 total flight hours. Toxicological tests revealed 0.021 ug/ml chlorpheniramine and 0.098 ug/ml diphenhydramine in the blood specimen. The therapeutic levels are 0.016–0.070 and 0.100 respectively. Chlorpheniramine and diphenhydramine are active ingredients in many nonprescription cold and allergy medications, and can cause drowsiness.

This private pilot flying a Maule M-5-235C was killed. The medications that were found in the pilot's body were common over-the-counter drugs. The level of the medications in the body indicated that the pilot had taken the proper dosage. This was not an "overdose" or improper use of the medication. The pilot simply had a head cold or an allergy problem. He probably took that same medicine for his symptoms all the time without concern. But when a medication gets into the bloodstream, it takes up space and limits the amount of oxygen that can be carried by the blood. Then if you climb to an altitude where there is less air, then it is a double-negative effect.

NTSB Number: ATL97FA029. Roxboro, North Carolina

Shortly after takeoff in instrument meteorological conditions (IMC), the pilot made an initial radio call to approach control (ATC), then about 35 seconds later, he announced, "I've just lost my vacuum." Air traffic controllers attempted to assist the pilot, but were unable to communicate with him further. Radar data showed the airplane made a left turn and climbed to 1,400 feet. After about 270 degrees of turn, the airplane's flight path became erratic. The airplane was last depicted at 1,300 feet and 2 miles south of the airport. About 4 minutes after takeoff, it collided with wooded terrain in uncontrolled flight. During the investigation, no specific reason was found for the reported loss of vacuum; however, during impact, extensive damage of the aircraft occurred. Toxicology tests of the pilot's liver showed 0.345

mcg/ml codeine, 0.134 mcg/ml dextromethorphan, 0.249 mcg/ml dextrorphan, and an undetermined amount of morphine. A test of the pilot's blood indicated undetermined quantities of opiates, a class of drugs that includes codeine and morphine. The findings of these substances is consistent with prior ingestion of codeine-containing cough syrup. Codeine, a narcotic pain reliever and cough suppressant, can produce impaired judgment, disorientation, and delayed reactions.

Probable Cause:

Failure of the pilot to maintain control of the airplane, due to spatial disorientation. Factors related to the accident were: the pilot's use of an unapproved drug, low ceiling, restricted visibility (fog), and an undetermined anomaly with the vacuum system.

A vacuum system failure (Chapter 3) is enough of a problem, but when the pilot has taken cough syrup that triggers disorientation, an accident is sure to follow. This was a private pilot flying a Money M20C. The pilot probably was not aware that something as seemingly harmless as cough syrup could prove to be deadly. Mixed with flying, this medicine became more dangerous than the ailment it was supposed to cure. The best flying medicine is no medicine at all.

Prescription Drugs

There is a "hit" list of drugs that a person cannot take and simultaneously hold an FAA medical certificate. The list is always changing as new drugs are tested and brought to market. If you are taking a prescription drug now and want to begin flying, you will want to discuss the use of that drug with an FAA medical examiner. If you are already a pilot and hold a medical certificate, then be careful what new prescription drugs that you take. I have had students who were prescribed medicine that if taken would have voided their medical certificates. In some cases an alternate medication that is not on the "hit" list might be prescribed.

I work with college students and have discovered that as a child several were given the drug Ritalin. Individual cases may vary, but the students I have worked with were denied FAA medical certificates until they had been off the drug for 10 years. Some of these had been on Ritalin until age 10 or 12, making it impossible for them to become pilots during most of their college years. Their parents did not know of this 10-year rule and had deep regret that their child had been on that drug when they found out. I am certainly not giving medical advice here, but you should consult an FAA medical physician about any medications.

NTSB Number: FTW95FA106. Arnaudville, Louisiana

The passenger reported that, while circling near a friend's house, he heard a "beeping" sound, and the airplane started "dropping quick." A witness stated, "I heard the engine rev, it looked as though the plane was trying to pull up, but it crashed into the tree and glided into the water and sank quickly." An examination of the airplane and engine did not reveal any preimpact mechanical abnormalities. Toxicology tests of the pilot showed that his blood contained 0.289 mg/ml of diazepam (valium) and 0.364 mg/ml of nordiazepam (metabolite of Diazapam). Valium is an antianxiety agent and muscle relaxant, not approved for use while flying. The drug concentration was in the therapeutic range.

Probable Cause:

The pilot's failure to maintain adequate airspeed while circling at low altitude, which resulted in an inadvertent stall and subsequent collision with terrain. A factor related to the accident was: the pilot's use of a drug that was not approved for use while flying.

This Cessna 150 pilot was killed and the passenger had a minor injury. Again, the amount of the drug present in the body was at a "therapeutic" level, which means the pilot had taken a recommended dose, but no dose was recommended when combined with flight.

Alcohol

I was surprised when I saw that there was even a single drunk flying accident. I thought that getting behind the wheel of a car while intoxicated is crazy, but getting behind the wheel of an airplane is insane. Surely a pilot, trained to be cautious and careful, would never do such a thing. I was wrong. Alcohol first attacks judgment and therefore takes away an ordinarily safe pilot's defenses.

Fortunately, this is a small category of accidents. Figure 10-1 has only 53 fatal accidents on the entire chart. There is good and bad news to these small numbers. It is good because there actually are fewer alcohol and drug accidents proportionally in airplanes than there are in automobiles. The average pilot may just be a safer person than the average driver. The bad news is that this number probably do not tell the entire story. These 53 accidents had alcohol or drugs listed as the primary cause of the accident. There have been accidents in other categories (stall/spin, landing, night flying), that could also have involved alcohol and/or drugs that were not classified primarily as an alcohol-drug related accident. Nevertheless, the Air Safety Foundation reports that less than one percent of general aviation accidents involve alcohol or drugs. Could you say the same thing for automobile accidents?

NTSB Number: BFO86FA021. Chesapeake, Virginia

Before the flight, the pilot was observed in a bar. He and three others then drove to the airport, where reportedly, the weather was bad and visibility was restricted by fog. After awhile, the pilot stated he was going to taxi the aircraft around the airport and invited the others to join him. Two of the others got into the aircraft, but the fourth person was concerned about the conditions and refused to go. The pilot started the engine and taxied to the end of the runway. The aircraft was then observed to take off and begin a turn to the right, then the witness lost sight of the airplane. Moments later, he heard the sounds of the plane crashing. An investigation revealed the aircraft had collided with trees and crashed on airport property, about 500 feet, east northeast, from the approach end of runway 28. No preimpact mechanical problem was evident. Toxicology checks showed the pilot had a blood/alcohol level of 0.08% and a urine/alcohol level of about 0.14%. The passengers had blood/alcohol levels of 0.03% and 0.07%.

All three people in the airplane were killed as the fourth person, who would not get into the airplane, watched. This was a private pilot with 123 flight hours in a Piper Cherokee 140. FAA regulations prohibit operation of aircraft above a 0.04% level. The pilot had a 0.08% level and was the most intoxicated of the three in the airplane.

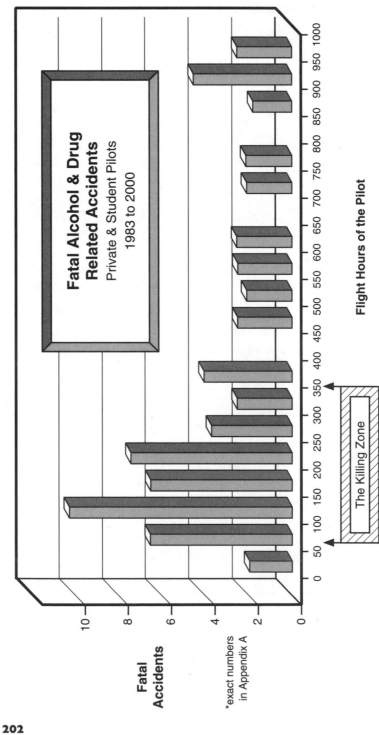

Fatal Alcohol & Drug Related Accidents
Private & Student Pilots
1983 to 2000

The Killing Zone

Flight Hours of the Pilot

Fatal
Accidents

*exact numbers
in Appendix A

Fig. 10.1

NTSB Number: BFO94FA134. Grand Lake Stream, Maine

The purpose of the flight was to give a sightseeing ride of Lake Pocumcus. While the flight was just above the surface of the lake, the pilot initiated a 90-degree climb. A witness stated that the airplane gained about 300 feet in altitude and lost "...almost all airspeed." The witness stated, "I believe the pilot pushed the nose over because the wings stayed level." The airplane descended straight down and impacted the water. Numerous witnesses stated they heard the engine running throughout the maneuver until impact. Postaccident examination of the airframe and engine revealed no anomalies. A photograph taken by a witness on the shore of the lake just prior to impact showed the airplane's elevators in a trailing edge up position. The pilot's blood/alcohol level was 0.075%.

Probable Cause:

The pilot attempted a maneuver at an altitude insufficient for recovery. A factor in the accident was pilot physical impairment due to alcohol consumption.

Both the private pilot and the sightseeing passenger were killed in the Piper PA-150. The photo of the airplane prior to impact with the water showing the elevator trailing edge up, indicates that the pilot was pulling back on the wheel. But if the airplane were stalled, pulling back on the wheel would have lead to a deeper stall and not to a recovery.

NTSB Number: MIA84FA042. Port Richie, Florida

The aircraft crashed into trees during a night forced landing after the engine quit. The surviving passenger said he had met the two pilots in a bar where they drank beer. They later went to the airport and flew across country about 70 miles to eat and have more beer. They took out beer for the return trip. During the flight back the engine quit, was restarted, and quit again. The passenger stated there was no fire because there was no fuel. No fuel was found in either tank. A cardboard box lined with plastic containing unmelted ice and beer was found on the back seat floor. The two pilot's blood alcohol levels were 0.217% and 0.386%.

This was the most blatant alcohol-related accident found. The FAA prohibits operation of aircraft above a 0.04% level. The 0.386% level of one of the pilots is in a category where unconsciousness and/or death due to respiratory paralysis is an effect (see table at the end of this chapter). This was a Piper PA-22 again with a student pilot/private pilot combination. It was unclear who was acting as pilot in command; however it was the student pilot with 75 hours that was in the left seat. That student's Class 3 medical certificate had expired. The rear seat passenger survived with serious injuries. This is the type of "accident" that really should not be called an accident. There were no victims here—only volunteers.

Alcohol is a "legal" drug, but it becomes illegal when combined with flying. The eight-hour "bottle-to-throttle" rule is also sometimes inadequate. Many factors combine to produce the effect that alcohol has on the body—and altitude is one of these. Alcohol is absorbed into the blood's hemoglobin faster than oxygen, so it's alcohol, not oxygen, that makes it to the brain first. This is what causes the drunk feeling—the absence of oxygen in the brain. This is called histotoxic hypoxia. Then when you take an already oxygen-starved brain to an altitude where there is even less oxygen, the impairment will increase. Each 10,000 feet of altitude doubles the effects of alcohol on the body. If you can't think, you can't fly.

Marijuana, Methamphetamine, and Other Drugs

There is a certain percentage of our society that uses illegal drugs, so I guess it should not be a big shock that a certain percentage of pilots also use drugs.

NTSB Number: ATL95FA088. Atlanta, Georgia

The non-instrument-rated private pilot received a weather briefing before departing. During the briefing, he was advised of approaching rain and thunderstorm activity. While taxiing for takeoff, tower personnel asked him if he was aware of a sigmet that was in effect for the Atlanta area; the pilot replied that he was. After takeoff, he departed northeast bound on a VFR flight to Raleigh, North Carolina. Witnesses at an electrical plant heard

a small aircraft flying in the area before they heard the aircraft hit the plant's concrete smoke stack. They then observed the aircraft descending out of the base of the clouds. The left wing was folded back against the fuselage, and the aircraft was spinning as it crashed into the ground. The smoke stack was about 880 feet tall, and its top was obscured by clouds. Impact marks were found about 75 feet from the top of the smoke stack. A toxicology test of the pilot's blood revealed 0.005 mcg/ml of tetrahydrocannabinol and 0.011 mcg/ml of tetrahydrocannabinol carboxylic acid (metabolites of marijuana).

Probable Cause:

The pilot's impairment of judgment and performance due to a drug, VFR flight into Instrument Meteorological Conditions, and failure to remain clear of the obstacle (towering smoke stack).

At least the pilot of this Cessna 172N was alone. The flight took place during the daytime. The private pilot was killed.

NTSB Number: LAX92FA112. North Palm Springs, California

The student pilot buzzed the location where he worked, and while maneuvering between 100 and 500 feet AGL made a series of 45–60 degree bank turns. A pilot witness reported this activity lasted for 5 to 10 minutes until the airplane abruptly pitched up, stalled, and descended in a near vertical nose down attitude until impacting the desert terrain. The pilot's blood alcohol level was 0.12%. In addition, Meth Amphetamine and Amphetamine were detected in the pilot's blood sample.

Probable Cause:

The pilot's failure to maintain airspeed during low-altitude maneuvers as a result of his physical impairment due to alcohol and drugs. Factors that contributed to the accident were: The pilot's poor judgment in buzzing his work location, and his failure to maintain adequate altitude.

This was a student pilot flying with an expired class 3 medical certificate and with a nonpilot passenger. The Cessna 150 pilot had mixed alcohol and drugs and as a result killed himself and the passenger.

NTSB Number: LAX96FA223. Burmuda Springs, California

A witness said the aircraft landed on runway 28, then took off again. After lift-off the aircraft was observed to accelerate over the runway, then climb in an extreme nose high attitude. While still in a climb, it suddenly rolled to the right and pitched forward. The aircraft continued to a nose low, steep descent until it disappeared from view. A flight instructor observed the pilot on the previous takeoff. He said that on that flight, the pilot pulled the aircraft's nose up until the aircraft almost stalled, then he leveled the aircraft. The flight instructor warned the pilot, but said the pilot replied that he had only slowed to 50 knots. The pilot was scheduled to surrender his pilot's certificate for a 120-day suspension following a second conviction for motor vehicle violations (DUI). A toxicology test of the pilot's blood showed 0.113% ethanol, 0.862 mcg/ml Meth Amphetamine (stimulant), 0.032 mcg/ml Amphetamine (stimulant), 0.005 mcg/ml Tetrahydrocannabinol carboxlic acid (metabolite of marijuana), and an unreported level of Tetrahydrocannabinol (marijuana). Tests also showed the pilot had 0.164% ethanol in urine and 0.143% ethanol in vitreous fluid. A test of the passenger's blood showed 0.009 mcg/ml of tetrahydrocannabinol, 0.028 mcg/ml of Tetrahydrocannabinol carboxlic acid, 0.878 mcg/ml Meth Amphetamine, and 0.092 mcg/ml Amphetamine.

Probable Cause:

The pilot's impairment of judgment and performance due to alcohol and drugs. (Meth Amphetamine, Amphetamine, and Marijuana), his excessive maneuvering (pull-up) after making a touch-and-go landing, and his failure to maintain sufficient airspeed, which resulted in a stall and subsequent collision with the terrain.

This was more than an accident; this was criminal activity. Some might call it a murder/suicide as the pilot killed himself and two others. This Cessna 172 private pilot with 110 hours had combined marijuana, alcohol, and methamphetamine on a "farewell" flight prior to giving up his pilot privileges to a suspension from two drunk-driving convictions. It is too bad someone didn't get to this person sooner, but he is off the road and out of the sky.

Alcohol and drug-related accidents are a small number of the total, but once again there is a Killing Zone present inside these numbers. Sixty-two percent of all alcohol and drug-related accidents involved pilots with greater than 50 but less than 350 flight hours.

The interaction of alcohol and drugs with flying is a slippery slope. When not under the influence of either alcohol or drugs, pilots think clearly and would never consider flying when impaired. But after partaking of alcohol and/or drugs, their normal, rational thinking disappears. When first robbed of good judgment, they can talk themselves into believing that they can handle it. But the very fact that they are entertaining the idea of flight is a symptom that the alcohol and/or drug already has an effect on them.

The FAA's Civil Aeromedical Institute (CAMI) offers this list of performance losses caused by alcohol:

- Judgment (normal cautionary attitudes are lost)
- Speed and strength of muscular reflexes
- Inhibitions and worries lessen
- Skill reactions and coordination
- Insight into existing capabilities
- Comprehension, and fine attention
- Efficiency of eye movement and hearing
- Sense of responsibility
- Relevance of response
- Ability to see under dim light
- Memory and reasoning ability
- Altered perception of situation

The list contains judgment, eyesight, hearing, coordination, comprehension, responsibility, and situation awareness. These are all tenets of safe flying and each one is attacked by alcohol.

A person's individual physiology is different from others. The symptoms of alcohol and drug use will vary in severity, depending on several factors. The type and amount of food that is already in the stomach, a person's body weight, the degree of body dehydration, and the speed in which the substance is consumed all change the reaction. The liver has the ability to filter about one-third of an ounce of pure alcohol per hour.

The blood alcohol level was repeatedly used in toxicology of the bodies after an accident by the NTSB. The same blood alcohol percentages are used to determine "legal" alcohol consumption limits. The following list is provided by CAMI and refers to blood alcohol concentrations:

0.01–0.05% Average individual appears normal.

0.03–0.12% Mild euphoria, talkativeness, decreased inhibitions, decreased attention, impaired judgment, slow reactions.

0.09–0.25% Emotional instability, loss of critical judgment, impairment of memory and comprehension, decreased sensory response, mild muscular incoordination.

0.18–0.30% Confusion, dizziness, exaggerated emotions (anger, fear, grief), impaired visual perception, decreased pain sensation, impaired balance, staggering gait, slurred speech, moderate muscular incoordination.

0.27–0.40% Apathy, impaired consciousness, stupor, significantly decreased response to stimulation, severe muscular incoordination, inability to stand or walk, vomiting.

0.35–0.50% Unconsciousness, depressed or abolished reflexes, abnormal body temperature, coma, possible death from respiratory paralysis.

The Civil Aeromedical Institute also gives this list pertaining to over-the-counter medications and the flight environment:

Pain/Fever Relief

Alka-seltzer, Bayer Aspirin, Bufferin

Ringing in the ears, nausea, hyperventilation.

Increases the effect of blood thinners.

Acetaminophen Tylenol

Liver toxicity in large doses.

Ibuprofen—Advil, Motrin, Nuprin

Upset stomach, dizziness.

Increases the effect of blood thinners.

Cold/Flu Relief

Antihistamines—Actifed, Benadryl, Cheracol-plus, Contac, Chlortrimeton, Dimatap, Drixoral, Dristan, Nyquil, Sinarest, Sinutab

Sedation, dizziness, impairment of coordination, upset stomach, blurring vision. Increases the sedative effects of other medications.

Decongestants—Afrin Nasal Spray, Sine-Aid, Audafed

Excessive stimulation, dizziness. Aggravates high blood pressure, heart disease, and prostate problems.

Cough Suppressants—Benylin, Robitussin, Vicks Formula 44

Drowsiness, blurred vision, upset stomach. Increases the sedative effects of other medications.

Appetite Suppressants—Acutrim, Dexatrim-

Excessive stimulation, dizziness, headaches. Interference with high blood pressure medications.

Stimulants—Caffeine, coffee, tea, cola, chocolate

Excessive stimulation, tremors, headaches. Interference with high blood pressure medications.

Smoking

The effects of smoking are also hazardous to flying. There are both long-term and short-term effects of smoking. Long-term effects include diseases such as emphysema, heart conditions, and many types of cancer. The U.S. Public Health Service has reported that cigarette smokers are 20 times more likely than nonsmokers to die from cancer. The short-term effects impact flight safety as well. Carbon monoxide (CO) constitutes up to 2.5% of the volume of cigarette smoke and more in cigar smoke. If a pilot smokes three cigarettes at sea level, a blood saturation level of 4% CO can result. Carbon monoxide is soaked up by the blood's hemoglobin 250 times faster than oxygen, so more CO instead of O_2 (breathing oxygen) is delivered to the brain and tissues. This greatly reduces the smoking pilot's altitude tolerance and night vision.

The FAA instituted the "I'M SAFE" program to help pilots make a preflight assessment of their ability to fly safely. I'M SAFE stands for: Illness, medication, Stress, Alcohol, Fatigue, Emotion. If you can honestly say that none of these factors are affecting you before flight—then go fly. But any one of these can adversely affect the safety of your flight. We use aircraft checklists to insure safe operations all the time. Use I'M SAFE as your body's checklist.

CHAPTER ELEVEN

Night Flying

FLYING AT NIGHT is one of the most rewarding experiences that aviation offers. Night flying is beautiful, but it is a completely different challenge to the pilot than flight in the daylight. Like anything else, though, when you are not familiar with the night environment and do not have much experience it can be dangerous. The regulations give a hint that night flying requires more skill and practice. If you make three takeoffs and landings to a full stop at night within 90 days, then you are also current in the daytime. But that does not work the other way around. If you make all your takeoffs and landings during the day, they don't count toward night currency.

When you fly at night the airplane does not know it's dark. The aerodynamics of the aircraft are exactly the same as in the daylight. The control surfaces, flight instruments, and magnetic compass all work the same. The only thing that is different is the human body's reaction to the dark. Our eyes see differently at night, our oxygen intake is more critical, and there are many night illusions waiting to fool us. Mistakes are amplified at night. The margin for error is smaller. And practice and experience make a difference in overall safety. The pattern of Figure 11.1 is now quite familiar. Of the fatal accidents shown on this chart, 59% involved pilots with between 50 and 350 flight hours. The 50 to 350 slice of the numbers is smaller than the rest, but has more fatal accidents than all the rest put together, revealing yet another Killing Zone, this time at night.

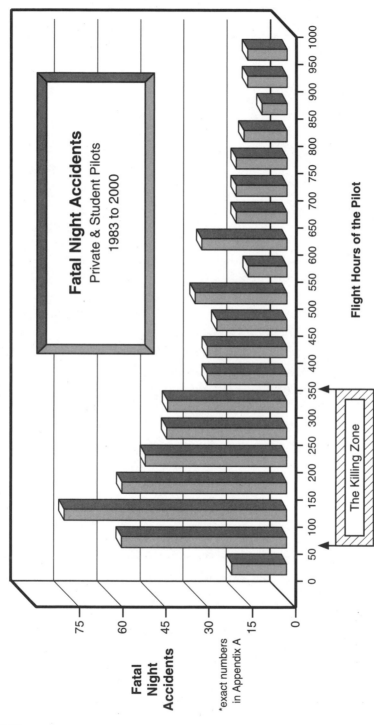

Fatal Night Accidents
Private & Student Pilots
1983 to 2000

The Killing Zone

Flight Hours of the Pilot

Fatal
Night
Accidents

*exact numbers
in Appendix A

75

60

45

30

15

0

0 50 100 150 200 250 300 350 400 450 500 550 600 650 700 750 800 850 900 950 1000

212

Fig. 11.1

In 1997 the requirements for the Private Pilot Certificate changed. The change included additional emphasis on night flight. Now the training must include three hours at night, which includes a night cross-country flight of at least 100 nautical miles round trip. The training also includes 10 takeoffs and landings at night with all landings made to a full stop. But no checkrides are given at night and there is no requirement that subsequent flight reviews are to be given at night. This means that those three hours of night training received as a student pilot must go a long way. Pilots may go their whole lives without ever actually being tested or evaluated at night.

Fuel reserve regulations are different at night. FAR 91.151 requires that the airplane must have enough fuel on board to fly to the destination airport plus 45 minutes. The reserve rule is 30 minutes during the day. The thinking here is that being low on fuel is more of a problem than during daylight hours. In the daytime you can see fields and open areas that could be used for an off-airport emergency landing. At night all those possibilities disappear. Recall that one of the accident examples from the fuel mismanagement chapter involved a pilot who was unable to switch fuel tanks and ultimately crashed. The accident took place in the dark, which made a successful forced landing unlikely.

The Air Safety Foundation reports that in 1998, 20.3% of all accidents were fatal and that accidents at night were more likely to involve a fatality than an accident during the day. "Twenty-four percent of the night VMC accidents were fatal as compared to 11.4% of the day accidents," Nall Report 1999. Then when night and IMC were combined, the percentages went up significantly. Sixty-eight percent of the night IMC accidents were fatal, making night-IMC three times more deadly than night-VMC accidents.

An examination of the fatal nighttime accidents exposed four scenarios where accidents happened:

1. Illusions when landing.
2. Dark night with no references in poor weather.
3. Dark night with no references in good weather.
4. Striking unseen objects on the ground.

Night Illusions

NTSB Number: SEA97FA215. Bremerton, Washington

While visiting friends, the private pilot arrived at an unfamiliar airport and requested the authorization to rent airplanes from an operator. The pilot intended on flying his friends on a cross-country night flight, but he did not inform the flight instructor who flew with him for the authorization. The instructor reviewed the pilot's logbook, but did not notice that it had been about 2 years since the pilot had logged any of his 3 hours of total night flying time. The operator's dispatcher subsequently allowed the pilot to rent the airplane for the flight, despite published night flying requirements which the pilot did not meet. During the return flight back to the airport, the pilot told controllers that he had difficulty locating the airport. During final approach, the airplane impacted trees and came to rest in a 3-foot-deep pond, about $1/2$ mile short of the runway. The Visual Approach Slope Indicator (VASI) lighting for the runway was out of service for routine maintenance, and a Notice to Airmen (NOTAM) had been issued. Examination of the wreckage did not reveal any evidence of preimpact mechanical malfunction. The front seat occupants, who were not wearing shoulder harnesses, drowned.

Probable Cause:

The pilot's misjudgment of distance and altitude from the runway, and his subsequent failure to maintain clearance from terrain. Factors contributing to the accident were dark night conditions, the pilot's improper decision to conduct flight at night, his lack of recent experience in the type of operation (night flying), his lack of familiarity with the geographical area, the inoperative airport visual approach slope indicator, and the improper dispatch of the airplane by the operator.

This was a private pilot with a total of 132 flight hours flying a Cessna 172. At night you cannot see everything, so the brain has a tendency to fill in the gaps. This leads to many types of misconceptions. Runways are required to have clear zones off each end. This is an area that is kept free of trees, power lines, or any other obstructions. This allows extra room if an airplane ran long off the end of the runway, or a landing came up short. But at night these clear zones become black holes. At larger airports these black holes are filled in with runway approach lights and guiding sequenced flashing strobes. At smaller airports the best you can get is a Visual Approach Slope Indicator (VASI) or the newer version, the Precision Approach Path Indicator (PAPI). On the night of this accident the VASI was out for maintenance, so all the pilot saw was the outlining runway lights beyond a large area of undefinable black. The black hole that must be flown across to get to most uncontrolled airport runways is scary. You cannot see where the clear zone ends and a tree line starts.

At night there is an illusion that can really be a trap. If you are accustomed to landing on a wide runway, a narrow runway at night can really trick a pilot. You know what the runway is supposed to look like on short final, but a runway that is more narrow than you are used to will never look the same on approach. It will always look as though you are higher than you actually are. A 50-foot-wide runway at 50-feet AGL can have the same proportions as a 100-foot-wide runway looks at 100-foot AGL. A narrow runway therefore will appear to be farther away because it's not grown in perspective to the size of a wider runway. This means you arrive at the ground sooner than you think you should. When airplanes crash short of a runway at night you must suspect this narrow runway illusion was at work. Also beware of twinkling lights. If a runway light seems to twinkle, that must mean that something else is passing between the light and your eye. That something else could be tree limbs or power lines. On final approach you want all lights to be steady, not twinkling. If you see a light twinkle—climb.

Our eyes have two kinds of "film," daytime and nighttime. Since humans are most adapted to the daylight, we have more daylight film called "cones." Cones are light-sensitive cells that line the inside back of the eye. Light passes through the lens in the front of the eye and is projected on the back of the eye and onto the cones. The cones send information to the brain through the optic nerve and together this creates daytime sight. In dim light, other cells are used called "rods." The back of the eye is actually a field of cone and rod cells that make up the retina. The daylight cells, the cones, are more concentrated in the center, where the incoming light is focused. This leaves the nighttime cells, the rods, out on the periphery. The eye attempts to focus images on the retina in the vicinity of the concentrated cones, but at night this becomes a problem. A problem because no nighttime rod cells are on the focus point. This is why at night dim lights can be better seen by looking to their side. This produces off-center viewing and the dim light lands away from the focus point and more on the periphery where the rod cells are. Lights can appear to move at night as the eye attempts to locate the light away from the focus point. Seeing lights that move when in fact they are stationary can produce an illusion called autokenesis.

The rods also need time to adapt. You know when you walk into a dark room from a bright room, you can hardly see anything at first. But as time goes by, vision in the dark room improves. This is because the rods produce a substance called Rhodopsin, or visual purple. It takes some time for this photochemical to coat the rods. But the moment the rods are exposed to bright light, the Rhodopsin degenerates and night vision is lost. This is why aircraft flight decks use red light, because red light does not destroy night vision. If you are flying at night and must use a white light, close one eye to preserve night vision in one eye. On dark nights, progressively turn down the illumination of cockpit lights so that the eyes can adapt further and outside vision will improve. But be careful when flying at night near thunderstorms. A flash of lightning will instantly steal your night vision. This is another good reason to stay away from thunderstorms.

Altitude has a direct effect on night vision. The brain is the number-one user of oxygen in the body. The eyes are an extension of the brain, so any reduction in oxygen will reduce vision. Without plenty of oxygen, the ocular muscles become weakened and uncoordinated. This causes blurred vision and difficulty in focusing. At night the effects are even worse because oxygen is required to produce Rhodopsin. The Civil Aeromedical Institute recommends that supplemental oxygen be used at night in unpressurized aircraft above 5000 feet. The regulation requiring supplemental oxygen does not lower at night, so it is legal to fly higher without oxygen, but it is not as safe.

Dark Night with No References in Bad Weather

NTSB Number: ATL92FA185. Samantha, Alabama

The pilot obtained a weather briefing and filed a VFR flight plan just before his departure shortly after midnight. Cloud layers

were forecast along the route of flight. **En route he contacted air traffic control and said he was lost. His conversation indicated that he was in clouds. Radar information depicted that the flight path of the airplane became erratic in heading and altitude. Subsequently, the airplane impacted the ground in a steep flight path and nose down attitude. The pilot had obtained his private pilot certificate about three months prior to the accident. His pilot's logbook indicated that he had been receiving instrument flight instruction. He had a total of 7 hours of simulated instrument flight.**

Probable Cause:

The pilot's inadequate preflight and preparation, lack of experience, which resulted in an encounter with instrument weather conditions during an attempted visual flight, and his subsequent spatial disorientation. Factors were the dark night conditions, clouds, and the pilot's geographical disorientation.

This flight had originated in St. Louis, Missouri, with an intended destination of Panama City, Florida, in a Piper 28-151. The private pilot who was attempting the trip had 118 flight hours. Flying into the clouds during the day is avoidable. Nevertheless, we saw examples of pilots who saw the clouds coming and with eyes wide open flew right into IMC. That VFR-into-IMC scenario is voluntary. At night you cannot always detect clouds before they are entered and so a VFR-into-IMC scenario at night can occur without warning. This is possibly why a VFR-into-IMC encounter at night is so deadly.

This accident report really hit close to home for me. When I was a private pilot with about the same number of flight hours as the pilot in this example, I also attempted a night VFR flight that encountered IMC. I was flying from Crestview, Florida, also with passengers, also in a Piper. Against much of my own advice throughout this book, I had flown to Destin and my buddies and I had spent the day on the beach. We had borrowed the airport courtesy car to get around, and as the sun set we drove back to the airport for the flight home. We took off at night and headed north. After leveling off for cruise flight, but less than 30 minutes into the flight, I suddenly noticed a red glow around the left wing and a green glow around the right wing. I had never seen this before and so at first I did not know what I was looking at. Then I realized that we had flown into a cloud and the glow was the position lights' reflection against the mist. I remember thinking about what I had been told about going into the clouds—I was sure we were going to die. I have never been so afraid in an airplane, but I made a standard-rate left turn. Within a few seconds the glow went away. Soon thereafter I saw the rotating beacon of the Crestview Airport and I flew straight for it. We were back on the ground, but my pals did not understand. They wanted to go home, and they wanted to know why we were back where we started. They never knew how close they came to being in an accident report. During the time of our short flight, the FBO manager had come by and locked up the courtesy car so we had no way to a hotel. We tried to sleep in the airplane, but it gets really cold, even in Florida at night. We ended up building a campfire on a concrete drainage ditch to avoid freezing. It was miserable, and my friends complained all night, but we lived to fly home the next day.

Dark Night with No References in Good Weather

NTSB Number: ATL94FA121. Youngstown, Florida

The private, noninstrument rated pilot, was returning home, VFR, at night. He initially received VFR flight following from Eglin approach, and was given a frequency change to Jacksonville Center. No subsequent communications were received from the pilot, although Jacksonville Center reported that the radio frequency that the pilot was given was not the correct one, and transmissions would not have been good on that frequency. Radar data indicated that the aircraft was traveling east-northeast at 2,400 feet when it entered a descending, right hand turn. Radar contact was lost at 400 feet, with the aircraft in a steep, right-hand descending turn. The wreckage was found in an open, plowed field. General disintegration of the wreckage indicated a high energy impact with terrain. No evidence of mechanical malfunction or failure was observed during the wreckage exam. The area of the crash site was remote farmland, with a marked absence of ground lighting. Local area thunderstorms and high cloud cover produced conditions conducive to spatial disorientation. The pilot's total flight time was 115 hours (12 hours at night).

Probable Cause:

Spatial disorientation experienced by the pilot, which resulted in his failure to maintain aircraft control. Factors were conditions conducive to spatial disorientation (dark night lighting conditions, high cloud cover, thunderstorms), and the pilot's lack of night flying experience.

The high cloud cover blocked out the stars and moon. The remote farmland gave no indication of terrain features or horizon. The pilot became suspended in a world with no visible up and no visible down. The pilot became disoriented without Instrument Meteorolgical Conditions. The visibility was reported to be 7 miles, but that does not help if there is nothing there to see. The pilot departed the Florida panhandle from Destin, in a Grumman AA-5B. The probable cause made no mention of the improper ATC handoff. The pilot used a radio frequency given to him by a controller that did not work. At the very least this would have been distracting and would have drawn the pilot's attention to the radios inside. Probably the pilot made several attempts to contact Jacksonville center, but nobody answered. The pilot must have tried to redial the frequency, and the airplane may have started the descending right turn when the pilot's attention was on the radio. When the pilot looked up, with no horizon, the pilot made an incorrect control input, one thing led to another, and control was lost. It is the same story as the John F. Kennedy, Jr. accident, except this happened over dark land instead of dark water.

NTSB Number: MIA89FA252. Ochopee, Florida

The pilot had received his private pilot certificate 5 days earlier. He took off with a friend at night for a flight over the everglades. When operations personnel realized the aircraft had not returned, a search was initiated. On 09-22-89 (two days later), the aircraft was found where it had crashed in the Everglades National Park. There was evidence that the aircraft had impacted in a slight nose down attitude. It was extensively damaged during impact and the main wreckage traveled about 120 yards before coming to rest. Both propeller blades were curled and contained chordwise scratches. No preimpact mechanical problem was found. The pilot had logged a total of 4.2 hours of night time flight.

Probable Cause:
Failure of the pilot to maintain attitude/clearance over the terrain. Related factors were: dark night, spatial disorientation of the pilot, and his lack of experience in the type of operation (night flying).

The accident investigators must have asked, "What is a person with a 5-day-old private pilot certificate, in a single-engine (Cessna 152) airplane doing flying over the Everglades at night!?" Good question. We see examples of pilots who have been given a privilege, but their lack of experience and poor decision making proves they were not ready to exercise the privilege. The visibility on the night of the accident was 15 miles, but there were overcast skies at 3000 feet. The Everglades has no ground lights. The overcast blocked any sky lights. The pilot flew with no references and eventually lost control.

NTSB Number: LAX89FA074. Yucca, Arizona

The noninstrument rated, recently certificated private pilot departed bullhead on Christmas Eve for a nighttime flight to Phoenix so his passenger could connect with an 8 o'clock airline flight home. The pilot took off and navigated to the southeast for 23 miles away from city lights and over desert terrain he became spatially disoriented and lost control of the aircraft. The aircraft descended at a high rate of speed and collided with terrain while in at least a 60-degree bank. There were no witnesses to the crash. Persons located 6 miles east of the crash site reported it was foggy, the visibility was 2 miles and there was light rain at the time of the accident.

Probable Cause:
Pilot's failure to maintain directional control. Contributing factors include the pilot's improper in-flight decision that was related to his overconfidence in his personal ability, spatial disorientation upon encountering inclement weather during the nighttime flight, and his total lack of night experience.

This is the third accident example that had all the same symptoms. The first over farmland, the second over a swamp, and the last over the desert. In each case there were no ground references and no horizon. The pilot in the last example was a private pilot with 108 hours flying a Piper 28-161. In each of these examples, and in the JFK, Jr. accident over water, there came a time in the flight where outside reference was unavailable. But in each case, inside reference was available. Even though technically VFR, in reality these were instrument reference flights flown by noninstrument pilots. This raises the obvious question: Should night flying require an instrument rating?

Several countries who follow International Civil Aviation Organization (ICAO) rules require that night flight be under instrument flight rules. ICAO rules themselves do not ban VFR at night, but do require more night training than FAA rules. Canada allows night VFR but requires a flight plan and a "night" rating. To obtain a Canadian night rating, the pilot must fly 10 hours at night—five dual instruction and five supervised solo. During the five dual hours, a two-hour cross country must be flown and 10 takeoff and landings made. In addition, 10 hours of instrument training is also required. In France, pilots must have a night endorsement from a flight instructor. Night-VFR cross-country flights must have a flight plan and can fly only along predetermined routes with specific reporting points. In several other countries, VFR at night is outright illegal. There is no movement underway in the United States to change the FAA's current night VFR rules.

Obstructions

NTSB Number: LAX94FA214. Hacienda Height, California

The helicopter collided with the upper static cable of power transmission lines while on a night cross-country flight. Witnesses reported seeing the helicopter flying low for several miles before the collision. After the collision, the helicopter came to rest on an interstate highway. Examination of the wreckage indicated the center post of the canopy struck the cable first. The visibility was reported as 6 statute miles with haze and fog. The wires are depicted on the helicopter route chart and the VFR terminal area chart. The cable support towers were lighted on the night of the accident. There was no evidence of preaccident mechanical failure or malfunction of the helicopter. Three automobiles were

damaged when the helicopter crashed. One automobile stopped in the westbound lane short of the crashed helicopter and another automobile struck the rear of the stopped vehicle. An automobile in the eastbound lanes struck the helicopter's mast and was damaged. All reported injuries on the ground were a result of automobiles colliding in the westbound lanes.

Probable Cause:

The pilot's selection of an en route cruise altitude that provided insufficient clearance, and his inadequate visual lookout. Factors in the night accident were reduced visibility as a result of fog and haze.

This accident involved a Robinson R-22 on a flight from Burbank to Chino, California. The pilot had a total of 122 flight hours, 66 of which were in rotorcraft. The pilot was a rated private helicopter pilot.

Night flying always helps make the distinction between a pilot that is "current" and one that is "proficient." Being current makes us legal, but being proficient makes us safe. Many pilots do not use a "personal minimum" when it comes to night flight. Just because a flight is legal, that does not make it smart. As a pilot you should be ready to cancel a flight whenever the conditions could be over your head, and be honest with yourself enough to know what is over your head. Your fly or no-fly decisions should be made based on your experience, proficiency, and comfort level, not based on what is strictly legal. There are times when VFR at night is without a doubt instrument flight. When planning a night flight, consider the terrain that must be crossed. We have seen that any place without ground reference (water, farmlands, swamps, deserts) can turn a VFR flight into an IFR flight even when clouds and visibilities are high. Stop these flights before they start, and look toward getting an instrument rating.

CHAPTER TWELVE

Ice

ICE IS NOT JUST a winter and northern concern. There have been ice-related accidents every month of the year and in all parts of the country. There are two main categories of ice accidents: carburetor and structural. Structural ice takes place when ice attaches to the outside structure of the airplane. It is usually associated with poor weather and IFR conditions, but carburetor ice can threaten the safety of any flight, even on warm, clear, VFR days.

Carburetor Ice

NTSB Number: ATL92FA111.
Lake Waccamaw, North Carolina

The pilot reported that the aircraft engine had a partial loss of power, and he was unable to maintain altitude. He stated that he applied carburetor heat, and the engine regained power. He stated that by the time the engine regained power, he was too low and the aircraft contacted the water. The temperature and dewpoint in the accident area at the time were conducive for the formation of carburetor ice.

Probable Cause:

The failure of the pilot to follow proper procedures while operating the aircraft in weather conditions favorable to carburetor icing.

This pilot had 130 total flight hours and was traveling from west to east from Lake Norman to Wilmington, North Carolina. This Cessna 152 pilot survived this accident, but his passenger was killed.

Ice can form inside the engine's carburetor even when the outside temperature is in the 90s. When ice does form, it does so in critical locations with the possibility of blocking the airflow to the engine. Without air, the air-fuel mixture cannot burn. If the air-fuel does not burn, then there can be no power stroke and the engine will quit. To remove ice or to prevent its formation, aircraft engines come with a carburetor heat system.

Figure 12.1 illustrates the simplest type of carburetor heat system. It is very efficient because it uses heat that the engine is already producing and would be wasted otherwise. The hot exhaust gases are collected and passed through a sleeve. The sleeve is more commonly referred to as a heater shroud. The hot exhaust gases pass through the shroud, through the exhaust pipe and out of the aircraft, and into the atmosphere. The exhaust pipe itself gets very hot, including the part of

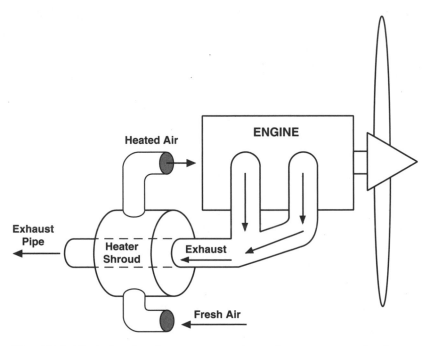

Fig. 12.1 Typical cabin heat system using exhaust gases and a heater shroud.

the pipe inside the shroud. The shroud is also full of air, but this is clean, outside air. The fresh air comes into the shroud and is heated by contact with the hot exhaust pipe. Then the fresh air that is now heated leaves the shroud and is ducted to the carburetor. The poisonous exhaust gases and the fresh air should never mix, only heat transfers. Inspection of the heater shroud should be routine to ensure that there are no cracks. A crack could allow exhaust to leak into the fresh air section. This is especially dangerous when a heater shroud is used for cabin heat. If a crack allowed exhaust gases into the fresh air on its way to the cabin, carbon monoxide poisoning of the pilot and passengers could result.

Figure 12.2 illustrates the carburetor and how either outside air or the heated air arrives at the venturi. On the left, ice is forming on the back of the throttle valve as outside air enters the carburetor. Outside air is made up of many gases from the atmosphere. Outside air also contains water vapor. Water vapor is H_2O in the gaseous state. A high water-vapor content will produce a high relative humidity and high dew point temperatures. The water vapor is carried into the carburetor

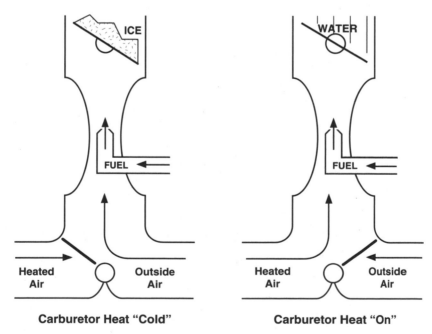

Fig. 12.2 The threat of carburetor ice.

along with the air. When the air accelerates through the venturi, the air's temperature drops because the molecules speed up and string out. This "stringing out" is called expansion. When the molecules are farther apart, there is less friction and therefore less heat. Acceleration makes the air cooler. When the air cools, it may very well reach the dew point, and fog will form inside the carburetor. A fog is a mass of liquid water droplets. So now air, fuel, and liquid water is traveling through the carburetor. When the air reaches the throttle valve, the air must once again accelerate to make it around the valve. As the air accelerates once more the temperature will drop even further, sometimes to a temperature below freezing. This is when ice can form, even when it is warm and sunny outside.

Figure 12.2 depicts the ice behind the throttle valve, but it can form along the interior walls of the venturi as well. The gap for the air to travel is already narrow, so any further narrowing of the venturi because of ice can choke off all airflow. As you can see, the ice threat is even greater when the engine is at low power. This is because the throttle valve will partially close to restrict airflow so the engine will go to a low RPM. When the throttle is pulled back to a low setting or idle, the air that does get through must pass through a slit between the rim of the throttle valve and the venturi wall. With the air passing through this small slit, it will take less ice to block the flow altogether. This is why most carbureted engines require carburetor heat during approach and landing.

Since the ice formation is a product of physical laws, we cannot stop its formation when conditions are right. But we can add heat to the carburetor in hopes that the temperature never drops to the dew point, or if ice does form we can add heat to melt the ice before it's too late.

The right side of Figure 12.2 illustrates the system with the carburetor heat on. The valve now prevents the cooler outside air from entering and clears the way for the heated air from the heater shroud to enter the venturi. If ice is present, the introduction of the heated air will momentarily make the engine run worse. This is because as the ice melts it is carried on into the engine, and water ingestion understandably will make the engine falter. But this will last only a short time and then the engine will run clear and smooth again.

To prevent carburetor ice and the threat of engine failure due to ice, pull on the carburetor heat whenever conditions are conducive for ice to form. I always operate with carburetor heat continuously on whenever I am flying in visible moisture: clouds, rain, etc. I also operate with continuous carburetor heat on when the temperature and dewpoint are within 5 degrees, or relative humidity is above 90%. Then, periodically, throughout every flight I apply carburetor heat and leave it on for a minute just to check. If you apply carburetor heat and the engine begins to run rough, then you did in fact accumulate carburetor ice. When this happens then you know that the conditions are right for ice, so leave the heat on. Consider changing altitude. A change in altitude will change the temperature and reduce the ice threat.

During the flight, any hesitation or engine roughness requires the carburetor heat to be applied. There are other problems unrelated to ice that could cause engine roughness, but you don't know what it is at first, so apply heat while there is still time. If heat is not applied and the engine quits due to ice, the carburetor heat will no longer work. Remember, it's the heat from the engine that we are using to heat the air. If the engine quits, it will no longer produce exhaust or heat. So when the engine runs rough you have a very small window of time to apply heat while you still have it.

NTSB Number: LAX94LA182. Catalina Island, California

After an overwater flight to Catalina Island, the pilot reported having engine trouble and that he was attempting to return to the Catalina Airport. The airport is located on a mountain top in an area of the island with steep mountainous terrain. A witness saw the aircraft and heard the engine repeatedly accelerate to full power, interrupted by sputtering and periods of silence. A second witness reported seeing the aircraft in a turn attempting to gain altitude as it approached the higher terrain. Ten minutes later, fire department units responded to reports of a brush fire and found the burning aircraft wreckage. Carburetor icing probability charts revealed the aircraft had flown in conditions conducive to moderate icing at cruise power and serious icing at glide power.

Probable Cause:

The failure of the pilot to apply carburetor heat as appropriate while conducting flight in conditions conducive for carburetor

**icing. A factor in the accident was the unsuitable steep moun-
tainous terrain for a forced landing.**

This private pilot, with 109 flight hours, was killed in the accident.
The airplane was a Cessna 152, which has a carburetor heat system
much like the one depicted in Figure 12.1.

It is a different problem when ice forms on the outside of the airplane.
Ice will form over the wings, on control surfaces, around antenna, on the
propeller, and covering air intakes. I witnessed a man taxi up to an FBO
who had just been caught in freezing rain. He had made an emergency
landing to get out of danger, and when he arrived on the flight line the ice
had completely covered his Mooney. We literally chipped him out
because ice had frozen the door shut. Ice had almost completely covered
the horizontal stabilizer and elevator combination. A few more minutes
in those conditions would have locked the elevator. The pilot did not
know which airport he was at, but he was very glad to be there.

Structural Ice

NTSB Number: SEA88FA034. Molalla, Oregon

**The recently certificated, noninstrument rated, private pilot flew
into an area of low ceilings, low visibility, rain, and snow without filing a flight plan. No record of a weather briefing was found. A surviving passenger reported that during flight, the pilot flew into
clouds that were initially scattered, but as the flight continued, the
clouds became thicker and darker. The passenger said they flew
into "ice clouds." Ice accumulated on the aircraft and the engine
began to falter. Reportedly, carburetor heat was applied too late
to prevent engine stoppage. A distress call was made at 1501
Pacific standard time. Soon thereafter, the pilot transmitted that
they were "going down" and then radar and radio contact were
lost. The aircraft collided with trees in mountainous terrain and
crashed. The emergency locator transmitter (ELT) did not operate. The aircraft was not found until 1012 PST the next day
approximately one mile from where radar contact was lost. There
was evidence that snow and ice had accumulated on the engine air
filter. Ice was found in the wire screen and plenum behind the filter. Carburetor heat was on. Residents approximately 5 miles**

away said heavy snow was falling. Two defective transistors were found in the ELT that would have prevented operation.

There were four people in this Cessna 182J. Only one person survived. The 129-hour pilot pulled on the carburetor heat, but ice in the carburetor was not the only problem. Ice covering the air filter would have blocked airflow to the engine. The air-fuel ratio is approximately 12:1, so for every one gallon of fuel burned, there must be 12 gallons of air passing through the filter. When the air could not get in, the engine started to suffocate. Why was a pilot, without an instrument rating, flying through the clouds? Even if he had an instrument rating, a Cessna 182 does not have the anti-ice and de-ice equipment to be flying in icing conditions.

Freezing rain is the greatest danger. Freezing rain occurs when rain in warm air falls through into subfreezing air below. Usually it gets colder, not warmer with altitude, so this condition happens when there is a temperature inversion. When a warm front moves across the surface, it will ride up and over a mass of more dense cold air. This can set up the inversion and trigger freezing rain. If you get caught in freezing rain you know that warmer air is above you, so one option is to climb to the warmer air. But the altitude of the warmer air may be higher than your airplane can climb. Your best bet is to land as soon as possible.

Ice on the airplane forms two different ways. In stratus clouds, rime ice will form. Rime ice looks like the frosty buildup in a freezer. Stratus clouds will most often produce light rain and drizzle. When the temperature is colder than the freezing point of water, these small droplets will freeze on contact with the airplane's surfaces. Rime ice is opaque because air gets trapped as the droplet freezes and one layer builds on the one below. Clear ice forms from larger drops and large drops come from turbulent air and thunderstorms. This means that clear ice will most likely form in cumulus clouds. Freezing rain will form clear ice and will coat the airplane surfaces with a smooth surface of ice.

Ice accumulation adds weight to the airplane, and this is a problem, but the real danger is that ice will change the shape of the airplane. The smooth cambered wing surfaces will become rough and ragged when rime ice forms. This disrupts the air flow and can destroy lift. At the very first sign of ice on the airplane, make plans to get out of those conditions. If working with ATC, tell the controllers that you have ice so that everyone can start working on the problem. If it is rime ice from stratus clouds, the good news is that the icing layer will not be very thick. Climb or descend and the problem will most likely be solved. If it is clear ice from cumulus clouds, then a diversion is needed. You probably cannot outclimb a building cumulus cloud to get away from ice, so go around it or back where you came from.

Before departure always look at the winds and temperatures aloft forecast. This will tell you if there are any temperature inversions. It will tell you where the freezing levels are. This information will help with the decision to climb or descend if icing is encountered.

NTSB Number: LAX93LA244. Eureka, Nevada

During the weather briefing for a cross-country flight, the FSS specialist told the pilot that VFR was not recommended. The briefing included precautions for moderate to severe icing; low ceilings; mountain obstructions; rain/snow; fog; and moderate to severe turbulence. The pilot asked about an alternate route and was told the same conditions existed on that route. About an hour after his FSS briefing, the pilot contacted FAA air traffic controllers for advisories. About two hours after his initial contact with ATC, the pilot reported he was in clouds and experiencing icing. The controllers gave the pilot a new heading to help get him out of the clouds. Several times the pilot told controllers he was in icing conditions. About ten minutes after first reporting icing and being in the clouds, the pilot said, "I am breaking out." No further radio contact was made with the pilot and the airplane disappeared from radar.

Probable Cause:

The pilot disregarding the advice that visual flight rules were not recommended; the pilot's inadequate weather evaluation along his route of flight; the pilot's attempting flight into known adverse weather conditions; the pilot's loss of aircraft control due to ice accumulation; and the pilot's limited actual or simulated instrument flight experience.

The probable cause cites the pilot's limited experience in actual instrument conditions. In fact the pilot was not instrument rated, but elected to fly anyway even though *"moderate to severe icing; low ceilings; mountain obscurations; rain/snow; fog; and moderate to severe turbulence"* was in his weather briefing. That is a "stay-at-home forecast" for even the most experienced IFR pilot. This pilot, however, flying a Beech K-35, did not heed the FSS warning and ultimately lost airplane control due to ice accumulation. The pilot was killed, but was flying alone.

NTSB Number: NYC91FA034. Perham, Maine

The pilot was told of forecast icing conditions during two weather briefings prior to departure. The pilot arrived at the destination airport and made 2 VOR approaches. During the approaches the freezing level was at the surface and there was a light drizzle. After the second missed approach the pilot requested vectors back to his originating airport. The controller said when last seen on radar, the airplane had descended from 2,000 feet to 1,200 feet in about 8 seconds. When found the airplane had large surface areas covered with $^1/_8$ inch ice. There was no evidence of ice on the surrounding terrain.

Probable Cause:

The pilot initiating flight into known icing conditions that resulted in an accumulation of airframe ice and a failure to maintain airspeed. The contributing factor was the existing icing conditions.

A light drizzle with below-freezing temperatures is sure to produce ice. When this Cessna 172 was discovered, it had ice on its surfaces but there was no ice on the ground. The obvious conclusion is that the ice formed on the airplane while in flight, not after the crash when on the ground. This was an instrument-rated private pilot with 251 hours of total time and 62 hours of instrument time. But a Cessna 172 is completely defenseless against ice.

Instrument Ice

NTSB Number: CHI89FA040. Carmel, Indiana

The aircraft departed for the cross-country flight in IMC conditions. While climbing to cruise, the pilot experienced a loss of aircraft control followed by an in-flight separation of both wings and

empennage. **Subsequent investigation revealed ice blocking the pitot/static tube. Icing forecasts were valid for the period covering the time of the flight and the pilot had received a briefing regarding possible ice. Although the pitot heat tested to be operational when checked during a post accident exam, the pitot heat switch was found in the off position after the accident.**

Probable Cause:

The pilot's poor understanding of the weather forecast, his failure to utilize the pitot heat in icing conditions, and the exceeding of the stress limits of the aircraft.

This general aviation accident was very similar to the Boeing 727 accident discussed in Chapter 3. The Piper PA-28-200 pilot had flown into known icing conditions but did not turn on the pitot tube heater. The pitot tube became blocked with ice and the airspeed indicator began giving incorrect readings. It is not known what other instruments, in addition to the airspeed indicator, may have also been affected by the ice accumulation. The incorrect readings led to loss of aircraft control. The pilot, in an attempt to recover but completely disoriented, tore the airplane apart in the air. This private pilot had 257 flight hours and was instrument rated. He and two passengers were killed in the accident.

To the thinking pilot, carburetor ice can be anticipated and prevented. Carburetor ice accidents still happen, however, because pilots get caught unaware of conditions that can produce the ice or are unfamiliar with how to get rid of the ice. The carburetor heat system is an excellent defense against the danger.

But there is very little defense for the light general aviation airplane against structural ice. Larger airplanes have systems to shed ice that accumulates, or to prevent it in the first place. Ice boots on the leading edge of wings will inflate and crack off ice. Heated windscreens will prevent ice from forming, making it possible for the pilot to see ahead. Electric heated propellers, stall warning indicators, and static ports all reduce or eliminate ice problems. But these features are not included on airplanes used for training and most personal flying. The only defense against known icing conditions in unprotected aircraft is just to not take off. If icing conditions are encountered while in flight, tell controllers, then plan to climb, descend, and/or divert. But do not hesitate. Take action to get out of the icing conditions immediately.

Killing Zone Survivor Stories

NASA Number: 418308

I was flying on a pleasure VFR flight. I was between layers of solid clouds. The layers converged. I pressed on because my destination was only 12 miles away with broken skies. The aircraft started to ice up and the pitot tube did too. All pitot-static instruments failed. I applied pitot heat and turned 180 degrees. I broke out into VMC in approximately 10 minutes. The airspeed, VSI, and altimeter functions soon returned. I returned home.

NASA Number: 361396

I was on an IFR flight plan from Vislalia, Louisiana, to Carthage, Texas. Despite the weather briefing that I had, that did not indicate any freezing rain, I began to ice up about 20 miles from my destination. I completed the NDB approach and got under the cloud layer, where it was only raining. The people at the airport wouldn't answer the unicom despite several calls. There was ice covering the windshield so I made a go-around. I was too low to receive Shreveport ATC. I didn't want to execute a missed approach because that would have put me back into the freezing rain. The ice finally melted off the windshield and I tried to land. There was a clear layer of ice on the wings that I couldn't see and when I tried to flare—it didn't. I landed hard and "tipped" the propeller blades.

CHAPTER THIRTEEN

The Effect of Advanced Flight Training

IN CHAPTERS 3 THROUGH 12 the accidents involving every phase of flight were discussed. But all the accident numbers used in those chapters were the accidents involving only student pilots, recreational pilots, and noninstrument rated private pilots. What about the accident rates of those pilots who go on to obtain an instrument rating on their private pilot certificate? Are these pilots safer?

Instrument Pilots

Figure 13.1 is the chart of fatal accidents among instrument-rated private pilots. In the early 1980s the instrument rating had a minimum time requirement of 200 hours. In 1986 that was changed to 125 hours. Then in 1997 the regulation changed again, this time requiring no specific amount of total time. Now an instrument rating requires the applicant to hold a private pilot certificate and meet all training requirements. This means that depending on how many hours it took to get a private certificate, a person could meet instrument rating requirements prior to 125 hours. This accounts for the first two flight hour segments of Figure 13.1 being zero. Between 0 and 99 flight hours there were no instrument-rated fatal accidents, but this is because there are no instrument-rated pilots with that amount of flight time.

Allowing for the fact that instrument-rated fatal accidents do not begin until the 100 to 149 range, another Killing Zone is evident. But this zone is shifted to a higher flight-hour range. The chart represents a total of 354 total fatal accidents involving instrument-rated private

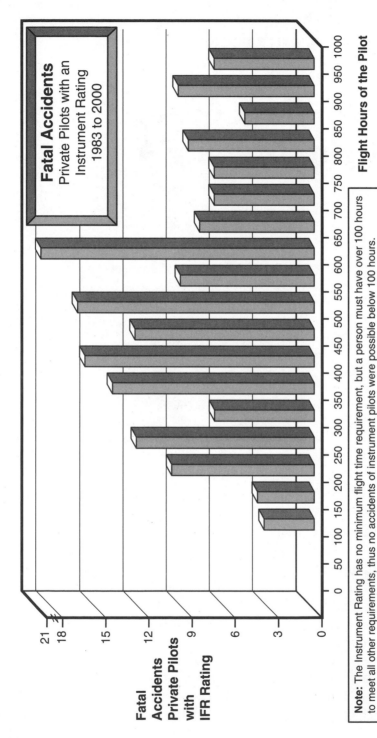

Fig. 13.1

236

pilots. Of this total, 48% of the accidents are concentrated among pilots who had between 250 and 550 flight hours. There again seems to be a span of flight time that has the greatest risk. These fatal accidents among instrument-rated pilots did not necessarily take place in instrument conditions. Many occurred in VFR conditions while on a VFR flight plan or no flight plan. Instrument-rated accidents that took place during instrument flight operations are discussed in Chapter 14.

Compare Figure 13.1 (and Appendix A14) with Figure 1.2 (and appendix A1). Figure 1.2 is the chart of all fatal accidents among student and noninstrument rated private pilots. At every flight hour level the instrument-rated pilots have far fewer accidents. The total student and noninstrument private pilots had a total of 2501 fatal accidents within their first 1000 flight hours. The instrument-rated private pilots had 365 fatal accidents between receiving the rating and reaching 1000 flight hours.

These numbers appear to show a huge safety improvement for pilots that get an instrument rating added on. Are these numbers a fair comparison? Maybe the reason there are fewer instrument-rated accidents is because there are fewer instrument-rated pilots.

It is true, there are fewer instrument private pilots than noninstrument private pilots. Figure 13.2 has the comparison number through the 1990s. You can see that of all the private pilots, approximately 20% or one in five have an instrument rating. But instrument-rated private pilots were involved in only 12.8% of the fatal accidents of private pilots. That is the same as saying that 80% of private pilots are not instrument rated, but noninstrument rated private pilots are involved in 87.2% of fatal accidents.

Taking the average number of private pilots each year from 1983 to 2000 and comparing them to the number of fatal accidents that private pilots had during the same time period, roughly 1 in every 120 noninstrument-rated private pilots were involved in a fatal accident. Over the same time period, roughly 1 out of every 150 instrument-rated private pilots were involved in a fatal accident. Clearly, pilots who go ahead and earn the instrument rating have fewer accidents.

There is an improving trend also evident in Figure 13.2. Every year of the 1990s saw increases in the relative number of instrument-rated private pilots. If instrument-rated private pilots have fewer accidents and if there are more and more instrument-rated private pilots around,

	Private Pilots	IFR Private Pilots	% IFR Private Pilots
1998	247,226	54,237	21.9%
1997	247,604	53,274	21.5%
1996	254,002	53,803	21.2%
1995	261,399	54,213	20.7%
1994	284,236	57,594	20.2%
1993	283,700	57,198	20.2%
1992	288,078	56,199	19.5%
1991	293,306	53,920	18.3%
1990	299,111	51,067	17.1%
	Source: FAA Airmen Registry		

Fig. 13.2 The rated 1990s had a percentage increase in instrument-rated private pilots.

then there should be fewer overall accidents. This may very well have been a leading factor that has driven the overall accident rate down.

Figure 13.3 is a comparison between IFR-rated private pilots and non-IFR-rated pilots in the category of nonfatal injury accidents. These exact numbers can be found in Appendix A15. Again, at every level the IFR private pilots had many fewer accidents. There were a total of 3060 injury accidents among pilots who had between 150 and 1000 flight hours. Of this total, 85.4% of the accidents involved non-IFR private pilots, meaning only 14.6% of the accidents involved IFR-rated private pilots. This also indicates that IFR private pilots are a safer group: 20% of the pilots are having only 14.6% of the nonfatal accidents.

Commercial Pilots

Part 141 flight schools can offer the training that leads to the commercial pilot certificate with a total of 190 hours of experience. This

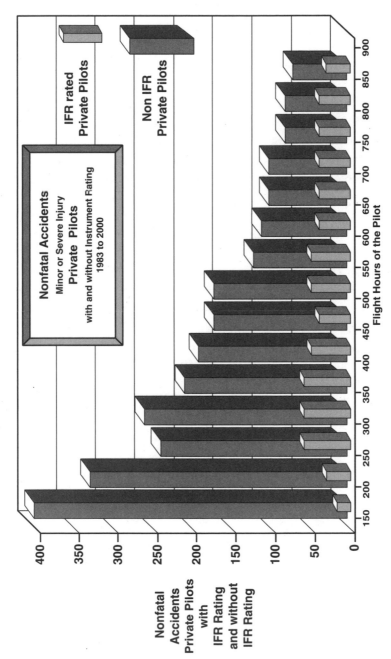

Nonfatal Accidents
Private Pilots
with
IFR Rating
and without
IFR Rating

IFR rated
Private Pilots

Non IFR
Private Pilots

Nonfatal Accidents
Minor or Severe Injury
Private Pilots
with and without Instrument Rating
1983 to 2000

Flight Hours of the Pilot

Fig. 13.3

experience may include a combination of flight and ground trainer time. Part 61 requires a total of 250 hours of experience, of which up to 50 hours can be in a ground trainer. Either way, Part 141 or 61, are commercial pilots from 200 to 1000 flight hours safer than private pilots between 200 and 1000 hours? Figure 13.4 illustrates the fatal accidents among commercial pilots between the time they get their commercial certificate and 1000 hours total time. This chart represents a total of 440 fatal accidents. Appendix A16 has the exact numbers.

Now taking just the flight hour levels from 200 through 1000 and combining commercial and private pilots, there were a total of 2010 fatal accidents. Of these, 440 were commercial pilots (see Figure 13.4 and Appendix A16) and 1570 were private pilots (see Figure 1.2 and Appendix A1). This means that 78.1% of fatal accidents involve private pilots, and 21.9% involve commercial pilots.

Using just 1998, there were 122,053 commercial pilots in the United States, and 247,226 private pilots. So there are roughly twice as many private pilots as commercial pilots, but the private pilots are involved in over three and half times more fatal accidents. And remember, in this comparison flight hours for the private pilot and commercial pilot are equal. Appendix A16 compares a private pilot with 500 hours alongside a commercial pilot also with 500 hours, etc. The numbers are clear. Pilots with advanced pilot certificates have fewer accidents.

Let's look at two typical pilots soon after each earns private pilot certificates. They both have equal flight time, and they both spend the first 25 hours after their checkride taking friends flying and making day flights to eat lunch and just do some sightseeing.

One pilot then starts working toward an instrument rating, the other does not. In the first year the pilot working for IFR flies on a regular basis. He is learning many new things: NDB intercepts, DME arcs, void times, timed turns—it is a whole new world. The pilot is practicing to hold the IFR tolerances of ±100 feet of altitude, ±10 degree of heading, ±5 knots of airspeed. He notices that on the occasional VFR flight he takes for fun that his flying skills are much better now than when he took his private pilot checkride, and his radio work is better.

Meanwhile the pilot who is not working on an instrument rating has stayed active, but his flights were sporadic. He makes several weekend cross-country trips, but two of these were canceled because of IFR weather. He attends a pilot safety seminar on mountain flying, but

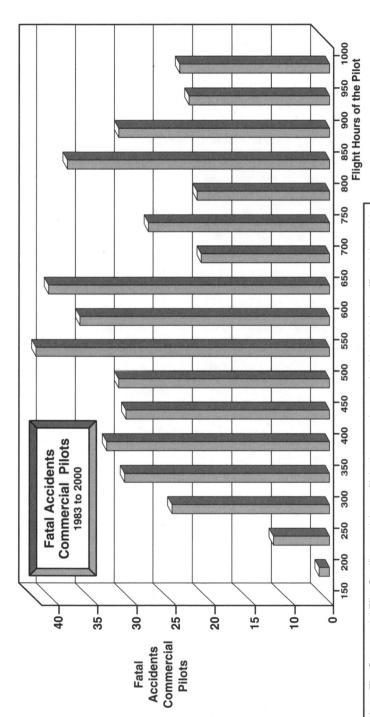

Note: The Commercial Pilot Certificate minimum flight time requirement is either 250 hours (Part 61) or 190 hours (Part 141) thus no accidents of Commercial pilots were possible below 190 hours.

Fig. I3.4

decides that mountain flying is over his head right now. In the first year since his private pilot checkride, he has made some memorable flights, but he is not night current anymore, and he has not ventured into any large airports. He knows that it would be tough for him to pass a private pilot checkride now. He would need a lot of brushing up.

The first pilot earns the instrument rating 13 months after his private certificate. In the next few months his goal is to stay instrument current and begin lowering his personal minimums. Since he has been active at his local airport and working with an instructor, he has heard about regulation changes that have come along; he feels pretty much up to date. He still takes some VFR trips, and he notices that his en route navigation is much better since he learned to navigate just with instruments. He never worries about getting lost anymore. Before the second year is over he checks out in a complex airplane and that starts him thinking about the commercial certificate.

The VFR pilot spends his second year making occasional local flights. He has given some rides to school kids, and flown to a football game in another state. But he now actively shies away from any large airports that would require rapid radio work. He really isn't sure anymore the procedure for radar control, and he is not confident about the airspace. He has not flown on a regular basis or with an instructor now for 15 months.

The first pilot is not interested in a career change to pilot, so why would he want a commercial pilot certificate? He recognizes that the training requirements for the commercial pilot certificate will push him to stay proficient and active. He feels very comfortable and knows the systems well on the complex airplane, and he loves to fly it. Soon he is learning even more flying—constant speed propellers, advanced maneuvers, and he is becoming a very wise decision maker. His flying skills have steadily improved. When he looks back at his proficiency two years earlier, he does not know how he ever passed the private checkride with as little as he knew then.

The second pilot enters his third year as a private pilot without any currency. He has not flown for over 100 days, but he wants to take his wife on a weekend trip to a state park that has an airfield. He stays away from the airport for the most part; he really does not want to be questioned about his lack of recent flying. If any regulation changes have taken place, he did not hear about them. His skills have eroded. There is no way he could meet private pilot standards on most maneuvers, and doesn't remember the procedure for many maneuvers.

The story of the two pilots is hypothetical, but as an instructor I know it's not far from the truth. Which of these two pilots would you want your family flying with? The pilot who kept climbing the ladder was exposed to new techniques, new ideas, current events, and was continually challenged. As his work to advanced ratings continued, his private pilot skills did not fade, they got better. If you can fly an ILS approach well, you can track a VOR great.

The FAA once completed a study entitled, "Private Pilot Flight Skill Retention 8, 16, and 24 Months Following Certification" (DOT/ FAACT-83/34). The object of the study was to determine how quickly a pilot loses proficiency. The FAA took a group of newly certificated private pilots and tested them again at 8-, 16-, and 24-month intervals just to see if their skills improved or eroded. All the pilots were at least up to private pilot standards at one time or they could not have passed their original checkride. Table 13.1 has the study results. On the left is a list of 27 tasks that private pilots must perform and meet minimum standards to pass the practical test. The first column of numbers is the percentage of pilots who met the standards on their original checkrides. You can then see the other columns are the percentages for the tests that were conducted at the 8-, 16-, and 24-month intervals. Comparing the numbers left to right is troubling at best. After 24 months' time, only 51% of the pilots could correctly enter an uncontrolled airport's traffic pattern and land or make a short-field landing. Only 38% could hold a specific rate of climb or properly execute a steep turn. Twenty-four months after the checkride only 33% remembered how to do compass turns (I can guarantee you if they had been working on an instrument rating they could have done magnetic compass turns!). The table proves what we already knew: If you don't use it, you lose it.

Table 13.2 ranks the tasks in the order of least retention. Most of these tasks look very familiar. Each is a cause factor in at least one category of accidents. Four of the tasks involve takeoff or landing. Accelerated stall and minimum controllable airspeed maneuvers are taught to prevent inadvertent stall. Remember how many accident examples have involved stalls or stall/spins. The loss of a vacuum system and its associated flight instruments have led to disorientation and accidents. When the gyro instruments fail, one of the backup instruments that must be

Table 13.1 Private Pilot Flight Skill Retention Project (DOT/ FAA/ CT-83/34).

Percent of Correctly Performed Flight Tasks			
Task	On Private Pilot Checkride	After 8 months	After 16 months	After 24 months
Takeoff checklist	100	98	100	94
Takeoff/departure	95	74	64	60
VOR tracking	79	68	48	50
Straight & level	72	74	76	65
Slow flight	83	62	37	39
Power on stall	99	77	79	71
Power off stall	98	84	80	76
Steep turns	79	54	51	38
Accelerated stall	90	51	52	57
Simulated engine out	92	88	67	77
Forced landing	95	74	67	76
Traffic pattern-uncntrl	89	70	52	56
Landing uncntrl airport	94	68	56	51
Short field takeoff	95	75	56	56
Short field landing	90	67	54	51
Soft field takeoff	94	80	65	61
Crosswind takeoff	93	89	53	75
Crosswind landing	93	81	58	63
S turns across a road	88	54	53	41
Turns about a point	83	52	52	41
Constant rate climb	84	56	62	38
Magnetic compass turns	74	51	40	33
Unusual att recovery - IR	97	66	70	66
180º turn - IR	90	79	63	52
Go-around	100	90	85	78
Landing cntrled airport	94	68	65	54
Communications	100	93	87	74

turned to is the magnetic compass, but magnetic compass turns is on the list of quickest skills lost. Where are the greatest number of midair collisions? In the traffic pattern and within 10 miles of an airport. Landing and traffic pattern at an uncontrolled airport are the two tasks with the greatest skill loss. I don't think it's a coincidence that the tasks in which pilots quickly lose proficiency are also the cause of many accidents.

When Bad Things Happen to Good Pilots

One of the biggest reasons that pilots who continue training beyond the private pilot certificate have fewer accidents is because they stay in a "safety mind-set." Pilots who never work with an instructor, who are always falling out of currency, and who do not stay up with current events, will become complacent faster. The pilot's personality (Chapter 15) is fed by new and different challenges, so if you deprive yourself of these challenges, your flying skills become dull. When you are not as sharp as you know you could be, you start an accident chain.

You know Murphy's Law: Whatever can go wrong, will go wrong. Well in practice Murphy is wrong. Usually things don't go wrong. An

Table 13.2

Flight Tasks with Greatest Skill Loss after 24 Months
1. Landing at an uncontrolled airport
2. Traffic pattern at an uncontrolled airport
3. Short field landing
4. Accelerated stall
5. Constant altitude turn
6. S Turn across a road
7. Turns about a point
8. Constant rate of climb
9. Magnetic Compass turns
10. Slow flight
11. Short field takeoff
12. Crosswind landing
13. Landing at a controlled airport
14. VOR tracking
15. Crosswind takeoff

accident does not occur every time a pilot is complacent. Most of the time when we forget a checklist item, or fly when we are not current, we get away with it. When we get away with it more than once, it becomes routine and soon our personal standards are lowered.

A normal preflight inspection includes a check of the fuel for contamination. You learned how to check the fuel as a student pilot and have been testing ever since. Then one day you were in a hurry and forgot to draw a fuel sample from the belly drain. The flight continued safely. You got away with it; after all you had tested the fuel from that drain 20 times before and had never found contamination. The next flight you reasoned that checking the drain was not a big deal; there was no problem last time and was never a problem before. In time the belly drain was completely dropped from your preflight routine. It became normal not to check it. Gradually what you considered to be normal shifted. Once you would not even consider starting the engine without first taking that fuel sample. That was normal then, but now you have a "new" normal. Your perception of what was acceptable crept over time. This "creeping normalcy" did not cause an accident for the next 100 flight hours, but someday...

This shift in what will pass as acceptable is not reserved for pilots. Air traffic controllers and maintenance technicians face the same problems. Furthermore it's not just aviation. I'm sure physicians, attorneys, teachers, police officers, and coaches all face the same threat.

I had an old car that I really liked once. It had a lot of miles and a lot of little things wrong with it. The automatic door locks did not work anymore, the carpet was worn, the horn started sounding funny, and it would not start unless you giggled the key a certain way. I looked past all these things. They did not break all on the same day or maybe it would have been more noticeable. I came to accept these things as normal about the car. My wife hated to ride in the thing because she saw its faults, I was used to its faults and after all I always seemed to get where I was going. My level of acceptance had shifted. I would have never purchased the car in that condition, but over time it didn't seem so bad. This is how good pilots get to a point of complacency without even knowing it.

How do we stop the shift? Go back to basics. Complete the entire checklist. Get a flight instructor to ride along from time to time. Fly with another pilot who has an advanced rating.

Airline pilots also fight this battle with creeping normalcy, but they have more factors to contend with. Because of bids and schedules it is very rare when the same pilots fly together for more than one trip. This means that total strangers are flying your airplane. How can people who have just met trust each other and perform as a team? Doesn't cohesiveness and interdependence develop over time? The airlines get around this problem by using standard operating procedures (SOPs). The airlines' SOPs are a strict set of guidelines that predefine roles and delegate tasks. The first officer, no matter which first officer it is, will always have certain responsibilities at certain times during the flight. The captain as well will routinely carry out jobs as appropriate. Even though they have never flown together before, they fly as a team because each person handles his or her share just at the right time. As long as all crew members follow SOP, then everything gets done.

General aviation has SOPs, but they are far less formal. We have checklists, operating procedures, recent experience rules, and practical test standards. We can fight creeping normalcy by protecting and maintaining our SOPs. In this way good habits will not slowly erode into bad practices.

Incentives

When I look back at myself when I was a private pilot, with flight time that would have placed me in the Killing Zone, I can't believe how or why I lived through it. I really did not know very much. At the time, however, I thought I did know enough to be safe. How naive, even foolish I was. Now I look at my flying in five-year increments. I know I have more experience and I am a better pilot today than I was five years ago. But five years ago I was not that bad. I am proud of my improvements.

Set a five-year plan for yourself. If you are a new private pilot, make a plan to be an instrument pilot with a 500-foot personal minimum in five years. If you are an instrument pilot, set your five-year goal to be a commercial pilot proficient in a complex airplane. Remember, the commercial certificate is not reserved for pilots who get paid to fly; it is for pilots who live to fly. If you are a private pilot you could also make a plan to become a seaplane pilot or a glider pilot within five years. If you really want to learn some flying basics and have a great time, push off the dock in a seaplane. If you really want to understand aerodynamics, you can see it firsthand when you let go of the tow line in a sailplane. You can even take advantage of the "new" airline transport pilot rules. Today, you do not need a first-class medical to become an ATP. You do not have to have 1500 hours of flight time to take the knowledge test. When you pass 1500 hours of total flight time, you can take the ATP checkride in a single-engine general aviation trainer-type airplane. The ATP is an instrument checkride combined with a systems test on whatever airplane you bring to the test. It really is not harder than another practical test. The ATP is not reserved for airline pilots. The ATP is for every pilot who wants to be at and prove that they are competent and growing pilots.

In five years you could fly a hot-air balloon, add a rotorcraft rating, and spend many hours at a radar room seeing what controllers do. And even if you take up none of these challenges within the next five years, at least be at level 5 of the FAA's pilot proficiency program in five years. The proficiency programs, sometimes called the "Wings" program, is one of the best deals going. You attend one FAA-sponsored safety program and fly three hours with an instructor each year. You get credit for a flight review, a reduction on insurance, and a set of wings to wear. Each year is a new level and a different set of wings. In some states the FAA teams up with the state's division of aviation to put on a "Wings Weekend." This event brings together volunteer flight instructors and safety program speakers for a 48-hour fly-in. Participants fly the three hours and attend a seminar all in one place over one weekend. I was a safety program speaker at a Wings Weekend once where over 500 wings were earned. That's 1500 flight hours in two days! Find out when your state will hold its next Wings Weekend and plan to attend.

The three hours of instruction that is required for the wings, whether you earn the wings in one weekend or over the course of a year, is broken down into parts. During the three hours, one hour is airwork (stalls, steep turns, slow flight, etc.), one hour is instrument work (instrument scan practice, instrument approaches, etc.), and the last hour is takeoffs and landings (crosswind, short and soft field). These areas of practice: Stalls, instrument work, takeoffs and landings, are what contribute to the most accidents. The wings program targets these areas because this is where the greatest prevention is needed.

The FAA should also provide incentives for pilots to remain proficient as they fly through the Killing Zone. At a minimum, the FAA should change the regulations so that private pilots without an instrument rating would be required to have an annual flight review until they have acquired 400 flight hours. After being instrument rated or after 400 hours, they can go back to the biannual flight review requirement. In this way private pilots will either move forward and become instrument rated or have at least one yearly instruction session while they are in the zone.

If you do not set a goal to improve your skills, then five years from now you will have lost ground. You will not be as good as you are today. Most importantly, you will be more dangerous to yourself, your family, and your friends.

This is not just a solitary decision. The actions of every pilot affect the rest of us. When accidents happen because pilots take off in bad weather, buzz their friends' houses, or fly under the influence of alcohol or drugs, it perpetuates the image of small planes as unsafe and their pilots as thrill seekers. I want people to come out to the airport and let a pilot show them how great flying can be. But they will be less likely to make the drive if they think "those little airplanes" are dangerous. If you are a pilot, you have a responsibility that goes beyond your own personal decisions. When you become a pilot you must also commit to be a lifelong learner. Start today and make a five-year plan.

CHAPTER FOURTEEN

Instrument Flight and CFIT

I T WAS PROVEN in the last chapter that pilots who continue their flight education beyond the private pilot certificate are statistically safer pilots. If you look back at Figures 13.1 and 13.3, the first thing that is apparent is good news: The numbers are small. There are not hundreds of accidents at each flight experience level. Also, the accidents are not as concentrated through one zone. The accident amounts are spread more evenly across the chart. Generally, this means that a lack of experience is less a factor in accidents involving instrument-rated pilots. This fact should be obvious since, by virtue of their instrument training, these pilots have more experience. And it's not just more experience; the instrument rating does more than add hours to a log book. The instrument rating adds seasoning, system-savvy, and the need for precision. Who gains more ground on becoming a safe pilot, a 150-hour private pilot practicing IFR flight or a 150-hour private pilot "boring holes in the sky?"

Figure 14.1 is a chart with very few accidents on it—again, that is very good news. But when the numbers are small, we should not attempt to draw too many conclusions. This is a chart of all the fatal accidents that occurred to instrument-rated private pilots, when the visibility was three miles and less. Once again the accidents are spread out more evenly across the chart, not concentrated into a particular killing zone.

Even though the accident numbers here are thankfully smaller, individually each accident that did take place was a tragedy in which lessons can be learned. The accident examples used in this chapter fell into one

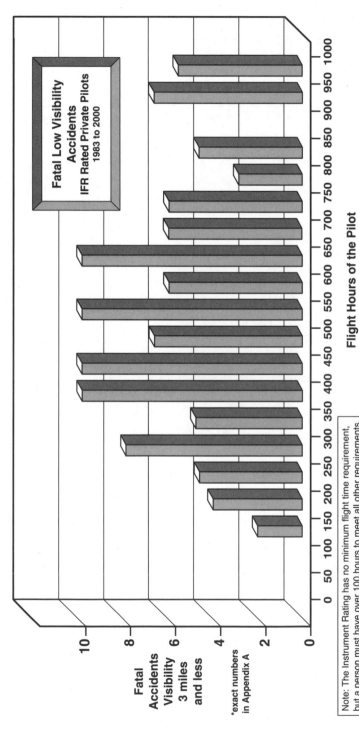

Note: The Instrument Rating has no minimum flight time requirement, but a person must have over 100 hours to meet all other requirements, thus no accidents of instrument pilots were possible below 100 hours.

Fig. 14.1

of two categories: Descent below minimum altitude followed by a CFIT accident, or an accident that violated the pilot's "personal minimums."

Descent Below Minimum Altitude and CFIT

CFIT is controlled flight into terrain. We have seen several accident examples, especially in Chapter 3, where the pilot involved lost control of the aircraft. This has happened when a pilot attempted VFR in IMC conditions, or flight with no outside references, followed by a loss of control while the aircraft was still in the air, followed by an out-of-control crash. The JFK, Jr. accident has become the most famous of that type. A CFIT accident is different. As the name says, a CFIT accident involves an airplane that flies, under control, into terrain or obstructions. At the time of the impact, the airplane was not out of control, the airplane was not stalled, the pilot was still flying, but the airplane was not in the right place at the right time.

NTSB Number: NYC84FA143. Concord, Massachusetts

The aircraft crashed into a house while on an ILS approach to runway 11 at Hanscom Field. Shortly before the aircraft had reached the outer marker (and after the hand off from Boston approach Control to Hanscom Tower) an altitude alert on the aircraft was received by the Boston Approach Radar Controller. He notified the Hanscom controller who then informed the pilot of the alert, to which the pilot replied he was at 1,400 feet and crossing the outer marker. This was the third instance during the flight where a discrepancy between the barometric altimeter and the encoding altimeter came to light. Investigation also revealed that the accident occurred at the end of an 18-hour work day for the pilot who had also been ill with the flu during the weekend preceding the accident.

The altimeter became an issue during the flight. Did the pilot just set the barometric pressure in the altimeter incorrectly? It would not be the first time an altimeter was set exactly 1000 feet off. When you start a flight, it may have been several days since the last flight of that airplane. During those days the pressure could have changed drastically.

Pilots see that the altimeter is not reading the proper field elevation, so they turn the knob and place the needle at the proper reading—but it's the 100s hand they set and they do not notice that the smaller 1000s hand is now either 1000 feet too high or too low. This error is never noticed and the flight begins. We do not know for sure that this was the cause of this particular accident, but the pilot did report at 1400 when in fact the airplane was much lower and eventually hit a residence. How is it possible to misread an altimeter? How can a pilot read 1400 when actually it is reading 400? It seems like a mistake that is so easy to catch. It is easy to catch when you are alert, following procedures, and thinking. It is precisely the kind of seemingly small thing that is easily missed when you have already been awake and at work for the past 18 hours and/or are recovering from the flu. The ceiling was 600 overcast, visibility was 1.5 miles, but most of all it took place in the dark. It would be possible for an airplane to descend below the clouds on a dark night and the pilot not be immediately aware that he or she is out of the clouds.

Most CFIT accidents that take place during an instrument approach occur on or near the final approach course. No matter what type of instrument approach is flown (ILS, VOR, NDB, GPS, etc.), there is always a straight line that must be flown that takes us to the runway. Very few CFIT accidents happen when a pilot strays off that straight line into terrain on either side. No, the instrument approach CFIT accidents usually happen along the straight line, but short of the airport when a pilot descends at the wrong time or flies below the charted minimum altitude.

NTSB Number: ATL95FA128. Raleigh, North Carolina

The pilot was unable to land at the Franklin County Airport, Louisburg, North Carolina, following two radar approaches and two instrument approaches due to low clouds and visibility. He then diverted to Raleigh-Durham airport for an Instrument Landing System (ILS) approach. According to recorded radar data, the flight path of the aircraft deviated from side to side of the localizer course centerline for most of the approach. The aircraft impacted the terrain approximately ½ mile northeast of the

approach end of runway 23L at approximately 400 feet MSL. The decision height for the ILS runway 23L approach is 636 feet MSL.
Probable Cause:
The pilot's failure to follow proper IFR procedures by descending below the decision height. Factors related to the accident were the weather conditions and the dark night.

This pilot, flying a Piper PA-28R-200, struggled by and basically stayed along the approach centerline. The report said the pilot deviated from side to side, but did not leave the approach. But this instrument-rated private pilot with 125 hours did fly below the approach's decision height, impacting the ground at 400 feet when 636 was as low as the approach allows. But this was also the second airport and fifth approach that the pilot had attempted; he was probably tired and desperate to get down. He may have been thinking that he had to make it in this time or not at all. The sky was obscured with virtually no vertical visibility. The forward visibility was reported as ½ mile. The weather being that low, no approach attempt could result in a safe outcome. It was not an approach decision that could have saved this pilot and his passenger; it was the takeoff decision.

Circle to Land

It is ironic, but one of the most dangerous parts of an IFR flight is the time spent outside the clouds when making a circle to land maneuver. Many airports have only a single approach. Since you only have one choice, often this means flying the approach with a tailwind toward an inactive runway. Once out of the clouds you must circle to a runway that is more favorable to the wind. The approach will have separate "circling" minimums that are higher than "straight-in" minimums but lower than a standard traffic pattern. During the circle you must keep the runway in sight at all times. If you fly back into a cloud or if the visibility is low and you get so far from the runway that it disappears, you must make an immediate missed approach. Not wanting to lose sight of the runway or run the risk of climbing back into the base of the cloud layer, pilots routinely fly low, tight circles. But tight turns, at slow airspeed, from low altitude are filled with hazard.

NTSB Number: CHI94FA089. Grain Valley, Missouri

The pilot flew two (nonprecision) instrument approaches to the destination airport. When he missed the first approach he reported he was having trouble and needed to make another approach. Witnesses reported everything seemed normal as the airplane overflew the airport while circling to land. They stated the airplane was about ½ mile northwest of the airport when the wings rocked back and forth, the airplane pitched nose up, then pitched down into the ground. The weather was reported to be 400 to 800 foot ceilings, 2 to 3 and one-half miles visibility in fog, and northerly winds at less that 10 knots. Records indicate the pilot had 9.6 hours in the accident airplane [Beech A36], including 5.1 hours actual instrument time and 3 approaches. There was no evidence of preimpact mechanical malfunction. The shoulder harnesses (front seat only) were in the stowed position.

Probable Cause:

The instrument-rated pilot's inadvertent stall while circling to land. The weather was a factor.

I have noticed that pilots let their guard down when it comes to a circle-to-land maneuver. All their energy and adrenaline is spent for the approach through the clouds. When they descend below the clouds and see the ground, they tend to relax a bit, because after all, VFR is easier than IFR. They feel relieved that the IFR portion is over and the VFR portion is no problem. In the pilot's mind, the flight is over, the job is done, the airplane is tied down. But in reality the upcoming circle-to-land maneuver in VFR conditions is more hazardous than was the approach through IFR conditions. The moment you relax your vigilance, an accident will be waiting.

Beyond Personal Minimums

The largest number of IFR accidents happen to pilots who get in over their heads. The IFR environment has much more variety than the VFR world. The VFR pilot's go/no-go decision is much more clearly defined: If the weather is IFR, I don't go. But when the weather is IFR, the IFR pilot does not automatically cancel the flight; no they must consider many degrees of IFR weather. Smooth stratus clouds, warmer than freezing, with a 1000 feet ceiling is great IFR. But drizzle, low visibility, and freezing temperatures is terrible IFR. The IFR pilot faces a much tougher decision than the VFR pilot. The decision is also easier for the VFR pilot because it is "illegal" to fly IFR when only VFR rated. This makes for a good reason when telling friends why they cannot fly and will now be inconvenienced. But when its IFR, the IFR pilot is legal to fly. But although legal, there are many IFR situations that are not safe. Pilots do not have the law as an excuse anymore. They must use their own judgment of the situation, and this means there will be times when a pilot can fly legally but not safely. Pilots with a low level of IFR experience do not handle this decision well. Some still have the rather immature attitude that, "If its legal, it must be Okay."

NTSB Number: BFO94FA103. Brainerd, Minnesota

The airplane was on an IFR flight plan. Prior to the approach, the pilot told ATC that he "needed all the help I can get, I'm kind of new at this." He was then cleared to descend and began receiving radar vectors for the ILS approach. The airplane began to head towards a nearby VOR after being vectored to the localizer course. The airplane intercepted the localizer and then disappeared from radar, radio communications were lost after the pilot declared a missed approach. The airplane impacted terrain south of the airport. The pilot had received his instrument rating 6 months before the accident. Several procedures specified in the ATC handbook were not followed by the controller. After the pilot initially began to stray from the localizer course, he was not informed of his position relative to the localizer, nor was an inquiry made as to the pilot's intentions. In addition, the vector provided to intercept the localizer course resulted in an intercept of 40 degrees; the ATC handbook stipulates a maximum intercept angle of 30 degrees.

Probable Cause:

The pilot's failure to follow the published missed approach procedure and his failure to obtain and/or maintain adequate altitude to avoid collision with the terrain. Factors in the accident were: The pilot's lack of instrument flying experience, and the

**controller's failure to follow standard procedures and to provide
more complete and helpful guidance to the inexperienced pilot
during the approach.**

The weather at the time of this accident was 500 overcast with rain
and fog. The private pilot had been an instrument-rated pilot for six
months. The pilot was completely legal for the flight, but the pilot got
in over his head. He said to the controller that he "needed all the help I
can get, I'm kind of new at this." There, in so many words, he said that
he was unprepared for the conditions and what lies just ahead. The pilot
had three friends with him in the Beech A36. Before takeoff he apparently was unable to say no to this flight even though more experienced
IFR pilots, facing that 500-foot ceiling, would have reconsidered.

When I was a student pilot I once flew into a large and busy airport
solo. My instructor told me to inform the air traffic controllers upon
my initial contact that I was a student pilot. By telling the controllers
that I was a student pilot, they were then aware to give me a little more
help, speak a little slower, and not put me in any tight spots. This lux-
ury is afforded student pilots, but there is no such thing as an instrument student pilot. Once you become instrument rated and start flying
in the system, you are expected to perform at the same level as everyone
else. Even if this is your first-ever flight solo in the clouds, the system
requires that you be as professional and as precise as any airline captain.
The step from student to private pilot is not near as tall as the step from
VFR to IFR. This is why there are accidents involving pilots who are
IFR rated but who are not ready for real IFR weather, or ready to work
in the real system.

The probable cause of the previous accident also aimed a finger at
the air traffic controller for not giving position information and assigning an intercept that was too steep. When you fly in the IFR system,
you must always evaluate the quality of the ATC assignments that you
receive. If you get an intercept that you feel is too steep, then ask for a
vector back around with a more shallow intercept to follow. You must
accept or reject your clearances. That being said, you will accept 95%
of them, but you must always be ready to question. Just like pilots, controllers get behind and make mistakes.

Can pilots ever blame controllers for their own lack of precision? I
think not. I think pilots must work through bad vectors, late intercepts,

forgotten EFC times, and slam dunks (controller term for a late descent resulting in a steep approach). I saw an interview once with a major-league baseball pitcher who had received the loss for a game. The interviewer said that he had pitched well enough to win, but took the loss because of three fielding errors. The interviewer gave the pitcher the opportunity to blame the loss on his teammates, but he would not. He said, "this is the major leagues, and in the major leagues you must pitch around errors and find a way to win anyway." Flying IFR in the ATC system is the major leagues. Just like that pitcher, when things don't go smoothly, when errors are made, IFR pilots can't assign blame; they must work around those errors. After the airplane is back on the ground you could call and talk to a controller about what happened, but in the air, pilots must take problems in stride; it's just part of the IFR game.

NTSB Number: LAX90FA031. Camarillo, California

During a night arrival, the pilot contacted the Naval Air Station at Point Mugu approach control and was cleared for a VOR runway 26 approach. As the pilot continued the approach, the aircraft veered radically off course to the right and descended prematurely. The controller issued a safety alert and told the pilot to turn left to a heading of 210 degrees and to climb and maintain 2,100 feet. The pilot acknowledged, but failed to take adequate corrective action. Subsequently, the aircraft collided with mountainous terrain at an elevation of about 1,100 feet, as it was approaching the final approach fix (FAF). Minimum descent altitude on that segment of the approach was 2,100 feet MSL. Approximately 10 miles southwest at Point Mugu (elevation 12 feet), the 2155 PST weather was in part: 1,000 feet overcast, 6 miles visibility with fog and haze. The pilot had been issued an instrument rating on April 28, 1989. Since that date, he had logged only 5.5 hours of instrument time and 4 approaches. [The accident occurred on November 12, 1989.]

Probable Cause:

Improper IFR procedure by the pilot, and his failure to maintain the minimum descent altitude for that segment of the VOR approach. Factors related to the accident were: darkness, low ceiling, fog, haze, the pilot's lack of recent instrument experience, and his probable spatial disorientation.

The pilot may have misread the chart or lost track of position along the approach. For whatever reason, the pilot descended before, not after, the final approach fix. The report specifically cites the pilot's recent experience. He had been instrument rated on April 28 and had completed 5.5 hours of instrument time and 4 approaches. Since the accident occurred in the month of November, the pilot was not legally current for IFR flight. The recent experience rules say that within the preceding six calendar months, a pilot must have six instrument approaches; this pilot had only four. At the time of this accident, the regulation also required six hours of instrument flight time. Since the time of this accident the FAA dropped the 6-hour requirement. But back in 1989, when this accident took place, the pilot was not legal to fly IFR. The pilot made a fatal mistake during the approach at a time when the law required him to have more practice making approaches.

NTSB Number: ATL93FA039. Phenix City, Alabama

The private pilot obtained her instrument rating three months prior to the accident. The first instrument landing system (ILS) approach into the airport terminated in a missed approach. When given a frequency change, just prior to reaching the outer marker, she replied that she was in serious trouble. The aircraft disappeared from radar, and no further radio transmissions were made by the pilot. Witnesses stated that the aircraft descended out of the base of the clouds in a nose down attitude, and a steep left bank. Examination of the aircraft revealed no mechanical nor instrumentation malfunction.

Probable Cause:

The pilot's failure to maintain control of the aircraft in instrument meteorological conditions. A factor was the pilot's lack of total experience in instrument flight conditions.

This pilot had 273 flight hours but only had been instrument rated three months. The weather at the time of her ILS approach was 300 overcast and 1.5 miles visibility. That is extremely low, but it was above the minimums for the approach. The fact that she made a missed approach tells us that the approach was not flown with precision; if it had been the airplane would have made it to clear air before the decision height with enough forward visibility to see the runway. The poorly flown approach forced a missed approach, and this prolonged the flight in IFR conditions. The pilot was given vectors for a second ILS attempt but got into "serious trouble" before the second approach began.

Why was a pilot, with three months since his or her instrument rating, out flying on a day with a 300-foot overcast? It was legal, the weather was above approach minimum, but it was below the pilot's "personal minimums." When you become an instrument-rated pilot you must work your way down. Your first solo IFR flight should be when the ceiling is greater than 1000 feet. Then as your time, practice, and experience goes up, your personal minimums can come down. Soon a 700-foot ceiling is within your comfort zone, and later even lower will be safe for you. But take it slow and one step at a time.

I canceled an IFR flight lesson once due to extremely low ceilings and visibilities and then ran into a newly instrument-rated pilot in the airport lobby. He had three friends with him and a pile of luggage. It was apparent that he was preparing to use his less-than-a-week-old instrument ticket to take his pals on a trip. The scenario looked just like one of these NTSB accident reports in the making, so I decided that I would stick around and see how long it took him to cancel this flight and make plans to drive. He called the FSS for a weather report and filed a flight plan. He checked out the rented airplane. I knew I had no authority to tell this pilot he could not go, but I decided I was going to do something to stop him if he kept going. Finally, he got up the courage to tell his friends the bad news. He canceled the flight and I never said a word to him.

NTSB Number: CHI96FA090. Brookston, Indiana

Prior to takeoff, the pilot called his girlfriend and asked her to pick him up at Gary Municipal Airport, Gary, Indiana, in one hour. One hour and 45 minutes later, the girlfriend contacted the Gary Air Traffic Control Tower, who advised her that they had no contact with the pilot. The wreckage was discovered the following morning. The pilot received a flight service station briefing prior to takeoff. The briefing included low ceilings and snow showers along the pilot's route of flight. A witness reported seeing "fast moving snow showers" in West Lafayette, Indiana, between 1900 and 1930 EST. Examination of the wreckage revealed no anomalies with the airplane. The pilot received his instrument rating one week prior to the accident. The pilot's logbook showed 2.8 hours of actual instrument time. The pilot's instrument instructor said the first time the pilot encountered actual instrument conditions, he became disoriented.

Probable Cause:

The pilot's disregard for the forecast weather conditions, his inadvertent flight into adverse weather conditions, and his disorientation. Factors related to this accident were the pilot's lack of experience in instrument conditions, the snow and fog.

This pilot's instrument rating was one week old. I wished someone had been there in that airport lobby ready to prevent this flight. Did this student's instrument flight instructor ever sit him down and have a talk about personal minimums? This pilot lacked experience in instrument flight, but more importantly lacked experience in decision making.

NTSB Number: MIA95FA023. Daytona Beach, Florida

The pilot was cleared by ATC for an ILS approach in instrument flight conditions. ATC terminated the approach when the airplane was observed on radar below the minimum altitude for that segment of the approach and "S" turning through the final approach course. The pilot was vectored and cleared for an ASR approach. The airplane was observed on radar turning through the final approach course, and the pilot was provided with heading corrections which were not acknowledged. The airplane continued turning through the approach course, and radar contact was lost. The pilot's last recorded instrument flight was on September 15, 1989, and his last recorded night flight was on February 18, 1989. [The accident took place on November 20, 1994.]

Probable Cause:

The pilot's failure to maintain aircraft control while maneuvering in instrument flight conditions during an instrument approach, resulting in the pilot becoming spatially disoriented, and a subsequent in-flight collision with terrain. Contributing to

the accident was the pilot's lack of recent experience in instrument and night flight conditions.

The instrument-rated private pilot was flying alone in a Piper PA-28-180. There are flying regulations that require pilots to practice. We have seen that following these regulations can make a pilot "current," but that does not guarantee the pilot is "proficient." The accident example is one of a pilot who is neither current nor proficient. His pilot records indicated that it had been over five years since he was current for either night flight or instrument flight; he was doing both at the time of this fatal accident.

The biggest challenge in all of flying is the transition from VFR flying to safe IFR flying. The accident examples in this chapter were drawn from a small group of fatal IFR accidents. The numbers show that the time spent working on and becoming an instrument pilot makes you a safer pilot. Part of becoming a safer pilot is staying current, proficient, and knowing your own personal minimums.

Killing Zone Survivor Story

NASA Number: 410713

The problem arose when I set 186 degrees in the OBS when 196 degrees should have been set. I was on a VOR runway 21 approach. Air traffic control gave me a heading to establish myself on the radial. I still had 186 degrees set on the OBS when 196 degrees should have been set. ATC cleared me for the approach and to maintain 2,500 feet until established on the radial. The needle centered, so I descended to 1,900 feet as shown on the approach chart. After 12.4 DME, I descended to 1,120 feet (the MDA). ATC then informed me that I was 1.5 miles east of the approach course. I initiated a missed approach. By this time I had realized my mistake and successfully completed a second approach.

A flight that descends in the wrong direction or descends too soon, but does not contact the terrain is called a CFTT—controlled flight toward terrain.

CHAPTER FIFTEEN

Advanced Aircraft Accidents

MANY OF THE ACCIDENTS that occur in advanced aircraft follow a similar pattern to other accidents. The same accident categories exist among pilots who fly aircraft with more power and more equipment. It is very ironic. You would think that a person who has become qualified to fly advanced airplanes would be an advanced decision maker. Sadly, however, this is not always true. Contained within the probable cause of accidents involving advanced airplanes is a high number of "overconfidence" and "lack of aircraft experience" statements. Pilots can get an airplane that they are simply not ready for. I have seen several pilots earn their private pilot certificate in a 100-horsepower trainer and then purchase a 200-horsepower airplane to use. The step up can be more than the pilot can handle.

The airplanes described in this chapter fall into three categories: complex, high performance, and multiengine. All the accident examples involve either student or private pilots.

A complex airplane is any airplane that has a constant speed propeller, retractable landing gear, and retractable flaps. This is the type of airplane that is specifically required for the commercial pilot and flight instructor practical tests. To act as pilot in command of a complex airplane, FAR 61.31(e) requires the pilot to have an instructor endorsement. There is no minimum training time that is required by the FAA, but most insurance companies will have a minimum checkout for coverage. In most cases the complex endorsement allows the pilot to fly a retractable-landing-gear airplane for the first time. The FAA Advisory

Circular AC61-65D, Appendix 1 has the endorsement: "I certify that (First name, middle initial, last name), (pilot certificate), (certificate number), has received the required training of 61.31(e) in a (make and model of complex airplane). I have determined that he/she is proficient in the operation and systems of a complex airplane." Any airplane that has the three components—constant speed propeller, retractable landing gear, and retractable flaps—is a complex airplane regardless of engine size.

A high-performance airplane is any airplane that has at least one engine with more than 200 horsepower. A light twin with 180-horsepower engines on each side is not high performance because neither engine is over 200. The total airplane horsepower (180 + 180) is not used to determine which airplanes are high performance. Regulation 61.31(f) stipulates that to act as pilot in command of a high-performance airplane, the pilot must have an instructor endorsement. Once again, Advisory Circular AC61-65D, Appendix 1 has the endorsement: "I certify that (First name, middle initial, last name), (pilot certificate), (certificate number), has received the required training of 61.31(f) in a (make and model of complex airplane). I have determined that he/she is proficient in the operation and systems of a high performance airplane."

You must be careful to get the correct endorsements after receiving your flight training in these advanced airplanes. I witnessed a very unhappy site when a commercial pilot flew across the state to take the flight instructor practical test with an FAA inspector. First on the agenda was to check the applicant's documents and endorsements. Sure enough, he had the wrong one. He had flown to the checkride in an airplane he was not endorsed to fly. Needless to say, he did not get his flight instructor certificate that day and had to retake his commercial checkride later.

An airplane can be neither complex or high performance, like a low-power, single-engine, fixed-pitch propeller, trainer. An airplane can be both complex and high performance, like a large twin. An airplane can be complex, but not high performance, like an airplane with a constant-speed propeller, retractable landing gear, and retractable flaps, but with less than 200 horsepower. And finally, an airplane can be high performance, and not complex, like an airplane with over 200 horsepower, but with fixed landing gear. Is a Learjet a complex airplane? No, because it does not have a propeller! A Learjet is probably a very complicated airplane, but it does not fix the definition of a complex airplane.

Figure 15.1 (and Appendix A17) is the chart that combines fatal accidents in both single-engine, high-performance airplanes and multiengine airplanes being flown by either student or private pilots. The accidents are not as concentrated as with previous charts, but there still is a Killing Zone. There are a total of 855 fatal accidents on the chart. Of these, 362 or 43% involved pilots with from 50 to 350 flight hours. Look over the following accident examples. These are advanced aircraft, but the same accident cause factors reappear.

VFR into IMC

NTSB Number: CHI93FA153. Kinsley, Kansas

The pilot told the airport manager that he had to return to his home base. The manager advised the pilot not to depart due to marginal weather conditions. The pilot asked the manager to obtain a weather briefing for him. The manager complied and again advised the pilot not to depart due to the weather. The manager stated that pilot was in a very agitated state, and told the airport manager he was going to depart. The manager advised the pilot to fly along the roads to avoid radio antennas between the departure and destination airports. The manager gave the pilot an old sectional chart because the pilot stated he did not have one. Shortly after the pilot departed, a witness observed an airplane of similar color and design pass low over her truck. She stated that there was a light mist at the time. The airplane collided with the ground right wing low.

Probable Cause:

The pilot initiated VFR flight into instrument meteorological conditions that resulted in a loss of aircraft control due to spatial disorientation. The adverse weather conditions were factors that contributed to the accident

The weather at the time of this accident was 300 overcast and 2 miles visibility. The pilot had one passenger aboard and both were killed. The airplane was a PA-28R-200.

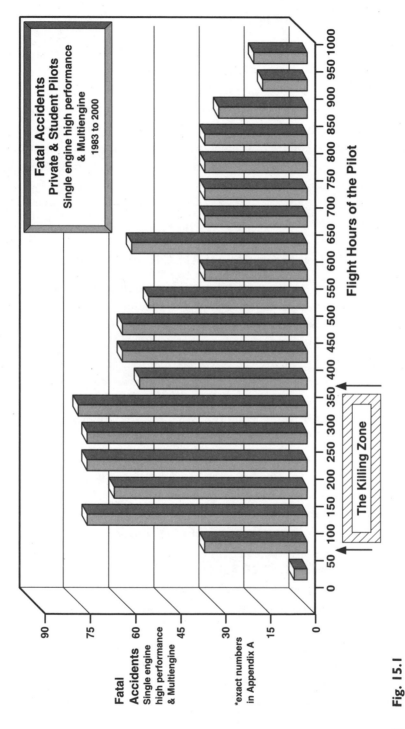

Fig. 15.1

Thunderstorm Penetration

NTSB Number: NYC91FA193. Fogelsville, Pennsylvania

The noninstrument rated pilot planned a VFR flight, but the weather at the departure airport remained IFR longer than anticipated. The weather was still marginal VFR, when the pilot took off. There were reports of turbulence, thunderstorms, and poor visibility in the area. While climbing through 6,700 feet, the radar data showed the airplane entered a rapid descent, with vertical velocity exceeding 10,000 feet per minute. The airplane wings separated in flight and a witness saw it spiral into the ground, followed by the wings 3 to 4 minutes later. The pilot had a total flight time of 218 hours.

Probable Cause:

The improper decision by the noninstrument rated pilot to attempt VFR flight in instrument meteorological conditions, that resulted in a loss of airplane control. Factors related to the accident were: The pilot's overconfidence in his ability, his lack of total flight experience, and his lack of an instrument rating.

Both the pilot and the single passenger were killed after the airplane entered a thunderstorm. The downdrafts inside a thunderstorm have unbelievable power. These downdrafts are responsible for windshears, microbursts, and gust fronts when they slam to the ground. No airplane, and not this Piper PA-28-236, can stay in one piece when hit by the full force of a thunderstorm. This private pilot had 218 total hours.

Takeoff

NTSB Number: SEA91FA110. Harrison, Idaho

During the initial climb from a private airstrip, the airplane was unable to climb over rising/wooded terrain. Subsequently, the airplane collided with trees and crashed. The runway sloped upward toward the rising terrain; a downwind departure would have been over descending terrain and a lake. Elevation of the airport was about 2,700 feet; elevation of the crash site (near the departure end of the runway) was about 3,000 feet. There were no reported mechanical failures or malfunctions.

Probable Cause:

Improper preflight planning/preparation by the pilot. The terrain condition and the pilot's lack of total experience were related factors.

This was a private pilot with 215 hours flying a Beech Bonanza 35. In most cases airports that have significant runway slope will require a land uphill/takeoff downhill procedure. Could this pilot have become overconfident in the abilities of this airplane in his attempt to take off uphill?

Lack of Currency

NTSB Number: CHI90FA236. Watertown, Wisconsin

The airplane collided with terrain approximately ½ mile south of the departure airport after a night takeoff. A police officer observed the airplane taxi to the runway; he did not see the airplane take off. He described the weather conditions at the time as heavy rain with lightning, and thunder, and high winds. A passenger in an automobile on a highway 1,500 feet west of the accident site reported what she thought to be a "shooting star." She stated they were driving slower than normal because of the heavy rain. The pilot stated to family members that he had to fly and do three takeoffs and landings before midnight or he would lose his license. His last flight was 92 days (exactly 3 calendar months) before the accident flight.

Probable Cause:

Initiation of VFR flight into adverse weather and instrument meteorological conditions. Factors that contributed to the accident were the adverse weather conditions and the pilot's self-induced pressure to maintain currency.

First let's review the regulations for remaining current. The regulation is 61.57, and it says, "no person may act as pilot in command of an aircraft carrying passengers, unless that person has made at least 3 takeoffs and three landings within the preceding 90 days." The report says that the pilot was concerned about "losing his license" if he did not fly before midnight. You cannot lose your certificate by losing currency. Maybe the family member quoted in the report misunderstood the pilot, but the pilot did not understand the regulation either. Calendar months have nothing to do with this regulation, so thinking that a flight by midnight was completely in error. Only a count of 90 days is at issue here, and the accident took place 92 days since the pilot's last flight. This means that the pilot was already noncurrent, and there was no need to hurry on the 92d day. Furthermore, the only privilege that you lose if you exceed the 90 days is the ability to carry passengers. Pilots could go out on their own, complete the landing and takeoff requirements alone, and regain currency on the spot. So the pilot was already unable to carry passengers, but because he misread or did not understand the regulation, he piled seven people in an airplane that has only six seats and took off into heavy rain, thunder, and lightning. All seven were killed.

The recent-experience regulations are in place because of a simple rule: Practice makes perfect. If you do not practice, your skills will disappear and at some point (90 days) the FAA feels that your skills have deteriorated to the point where it is not safe to take anybody else along. You can risk yourself, but not others past 90 days. Why did this pilot risk his life to meet the requirements of a safety rule? The private pilot was flying a Piper PA-32R-300 Lance. The pilot had 212 total hours, but only 8 in the Lance.

Weight and Balance

NTSB Number: ATL95FA142. Mount Dora, Florida

The pilot and passenger were en route to Oshkosh, Wisconsin, to attend the Experimental Aircraft Association annual meeting. They had intended to camp in tents while at the meeting. The pilot departed his home earlier in the morning, and flew to the Mid-Florida Airport to pick up a passenger. After arriving, the pilot told witnesses that during the flight, the airplane's center of gravity was aft, and he had to use a lot of forward trim to maintain level flight. The pilot and passenger hurriedly loaded the passenger's baggage

on board the airplane, without weighing the baggage, and attempted to depart. The takeoff roll was longer than normal, and the airplane climbed to about 300 feet. The airplane began a right turn, then entered a left bank, which continued to steepen. The airplane then entered a steep descent and crashed in a lake. Weight and balance calculations showed the airplane's gross weight (GW) was 2,765 pounds, and its center-of-gravity (CG) was 85.4 inches. Its maximum certificated GW for takeoff was 2,757.5 pounds, and its aft CG limit for that weight was 85.1 inches.

Probable Cause:

The pilot's inadequate preflight planning/preparation by allowing the airplane's weight and balance to be exceeded, and his failure to maintain control of the airplane, while turning after takeoff, which resulted in an inadvertent stall.

When pilots move up to more powerful and advanced airplanes, they can overestimate what the airplane can do. When you fly a low-performance airplane you don't expect much. You would never dream of overloading and then expect the airplane to perform. But the pilot of this Beech Bonanza 35F must have thought that the airplane's additional power could overcome the situation. Both the passenger and the private pilot with 212 hours were killed in the accident.

Multiengine

An instructor endorsement alone is not enough to earn multiengine flight privileges. To act as the pilot in command of a multiengine airplane, you pass a practical test. But there is more than one kind of multiengine practical test. First, you can become multiengine rated on a private pilot certificate. The accident examples in this chapter involve multiengine private pilots. If you are obtaining a private-multi, then the practical test will be given to the private pilot level. More commonly, pilots get a commercial-multi, and that practical test would be given to commercial standards. Then there are VFR and IFR multiengine practical tests. When the test begins, the applicant is supposed to state which multiengine test this is to be. Once that statement is made, there is no turning back. You cannot declare that it is an IFR-multi and later switch to VFR-multi when you do not meet the standards on just the

IFR portion of the test. Multiengine flying is quite demanding, and once again the accident reports point to the fact that pilots get into trouble because they are flying an airplane that they were not ready for.

No Multiengine Rating

NTSB Number: MIA85FA112. Jacksonville, Florida

The single-engine land, private pilot had an instrument rating, but had no rating to fly multiengine aircraft. While on an illegal flight with marijuana on board, he contacted Jacksonville Approach control and advised that he could not see the ground due to dense fog. The aircraft was low on fuel and he requested assistance in landing. The pilot was vectored for an ILS approach to the Jacksonville International Airport, but he was unable to complete the approach. The aircraft's radio equipment was old and provided poor communications, but ATC personnel managed to vector the aircraft for a PAR approach to the Naval Air Station. The first attempt was unsuccessful and the aircraft's fuel state became critical. Subsequently, an engine lost power from fuel exhaustion. While the pilot was attempting a single-engine approach, the aircraft entered into a right turn and crashed in a steep nose down, right wing low attitude. An exam of the wreckage revealed the right engine was feathered and left prop had damage consistent with high power.

The pilot of this twin-engine Piper Aztec had no multiengine rating; nevertheless, he was flying when the weather was ¼-mile visibility with 100 overcast due to fog. The fuel ran out on the right engine and eventually control was lost. A single-engine approach is required to pass the IFR-multipractical test, but this pilot had not passed that test. There were two persons onboard and both received fatal injuries.

Lack of Experience

NTSB Number: MIA 97FA020. New Bern, North Carolina

The airplane was over gross weight at takeoff but was within Weight and Balance at the time of the accident. Witnesses

observed the airplane flying low with the landing gear retracted over a wooded area then observed the airplane bank to the left and pitch down. The airplane then pitched nose up and entered what was described as a flat spin to the left. The airplane descended and impacted the ground upright with the landing gear retracted and the flaps symmetrically extended 6 degrees. Examination of the flight control systems, and engines revealed no evidence of preimpact failure or malfunction. A cabin door ajar indicating light was not illuminated at impact but the gear warning light was illuminated at impact. The pilot recently purchased the aircraft and accumulated a total of 1 hour 23 minutes during 6 training flights. He accumulated an additional 3 hours 37 minutes after completion of the training flights while flying with other qualified pilots. The accident flight was the first flight in the make and model while flying with no other multi-engine-rated pilot aboard.

Probable Cause:

The pilot's failure to maintain airspeed (Vmc). Contributing to the accident was his lack of total experience in that kind of airplane.

This was a Piper PA-60-600 Aerostar. The pilot held the private pilot certificate with 382 hours. This was just too much airplane for someone with these few flight hours. The probable cause statement called it "lack of total experience." All three people aboard were killed.

VFR into IMC

NTSB Number: DEN89FA218. Hesperia, California

The pilot and his wife departed apple valley airport at 0310 PDT on a flight to Riverside, California. At approximately 0320, two truck drivers saw the aircraft flying at low altitude toward Cajon Pass (elevation 4,120 feet MSL). Subsequently, the truck drivers saw the aircraft burst into flames when it crashed. Although the prevailing weather at Apple Valley and Riverside were reported to be VMC, the truck drivers said the weather at Cajon Pass was shrouded in fog and drizzle. The pilot was not instrument rated. Friends of the pilot and his wife said that he was concerned about the weather, but she was anxious to return to Riverside.

Probable Cause:

The noninstrument rated pilot continued VFR flight into instrument meteorological conditions (IMC) and misjudges his altitude above the terrain. Contributing factors were: The pilot's lack of instrument experience, pressure induced by his wife (the passenger) to take the flight, dark night, weather conditions, and mountainous terrain.

All the pressures that can affect a single-engine pilot's judgment are still present when flying a multiengine. This Cessna 310 was flying in 1/8-mile visibility due to fog. The pilot was a private pilot with a multiengine rating and 339 total time. The husband and wife lost their lives in this completely avoidable accident.

Takeoff

NTSB Number: CHI92FA257. Crystal, Minnesota

Witnesses stated as the airplane began its takeoff roll on runway 23, the gusty winds shifted to a direct right crosswind. They reported the airplane remained on the runway longer than usual, and the nose "wobbled" back and forth as it "skipped from main (landing gear) to main." One witness stated "the aircraft was pulled off right at the end of the runway and immediately banked to the left." The aircraft continued in a nose high, steep ("knife edge") left bank, then descended into trees and a residential community. Postimpact fire ensued. The multiengine rated, private pilot's logbook indicated 347 hours total flight time, including 253 hours in the accident make and model aircraft. Postaccident investigation revealed no evidence of preimpact mechanical anomaly. Witnesses reported "squirrely" winds and "squall line" type thunderstorms in the area at the time of the accident.

Probable Cause:
The pilot's failure to maintain aircraft control during takeoff/initial climb in gusty, shifting crosswind conditions.

Takeoff is always a critical phase of flight no matter what type of aircraft is involved. Then add gusty crosswinds and it can be too much. Remember, takeoffs are always optional. Three of the five aboard this Cessna 310 were killed in the accident.

Low Aerobatics

NTSB Number: LAX88FA264. Portola Valley, California

While dining with friends with whom he resided, the pilot told them that he was "going flying." At the completion of the flight he would fly over the house before he returned to Palo Alto airport (PAO). The pilot twice circled the residence which is about 7 miles southwest of PAO at a low altitude and then proceeded west. Ground witnesses reported that when the airplane was about ½ mile west of the residence it entered a 70-degree nose-high attitude. The maneuver resembled a left hammerhead stall except before it reached the nose-down attitude it entered a spin. One witness close to the accident site reported that the airplane had stopped its spin rotation and was in about a 45-degree down attitude at impact. The witnesses near the accident site reported that one of the engines sounded irregular. The postcrash investigation disclosed no evidence of any airplane or engine preexisting malfunctions or failures.

This fatal maneuver was performed in a Beech 76 Duchess. At one time NASA conducted spin tests on an experimental version of the Beech Duchess. The result was a prohibition against spinning in that and most all other light twins. This pilot was flying alone.

Personal Minimums

NTSB Number: MIA87FA062. Ft. Lauderdale, Florida

The private pilot had been cleared for 27R ILS approach and was on the localizer at 1,600 feet when the controller noticed the aircraft north of the localizer course and in a right turn headed east. The controller queried the pilot as to his heading and the pilot replied, "east, which was to go?" The flight was then given a turn so as to avoid another aircraft on the approach that the accident aircraft was approaching head-on. The pilot never acknowledged any further radio transmissions and shortly thereafter, the aircraft dropped off radar and crashed into the Atlantic Ocean in 800 feet of water. The newly rated multiengine pilot had been advised by his flight instructor not to conduct any night IFR flight into heavy IFR conditions until he gained more IFR experience. The weather conditions were night IFR with rain, heavy turbulence, and thunderstorms in the area. Neither the aircraft nor occupants were recovered from the ocean.

The instructor had warned the pilot about "personal minimums" in a roundabout way. This Piper PA-34 Seneca pilot ignored this advice and was flying with a passenger on a dark night with a 500-foot overcast ceiling.

Landing Gear

NTSB Number: FTW89LA135. Okmulgee, Oklahoma

The nose gear failed to fully retract after takeoff and ground personnel confirmed that it was not fully retracted or extended, approximately one-half mile from the runway the pilot elected to secure both engines. He said the aircraft immediately entered a steep descent. The aircraft impacted the terrain approximately 1,080 feet short of the runway.

Probable Cause:

Failure of the nose gear extension assembly and the pilot's improper precautionary landing procedures that included securing the engines prior to a point where a safe landing was assured.

This was not a fatal accident, but the pilot of this Cessna 421 was seriously injured. He was attempting to reduce the damage to the airplane, and that is admirable, but people should always come before airplanes. There was another multiengine pilot that could not get the nose gear down and locked. There were six passengers on board. In an attempt to minimize the airplane damage caused by a nose-gear-up landing, he had all the passengers move to the extreme rear of the airplane. The pilot hoped that this would shift the weight rearward, making it easier to keep the nose in the air until the slowest possible speed. The plan worked. The nose contacted the ground but only after the engines were secure and the forward speed was so slow that very little damage took place. So the airplane was saved, but the pilot lost his license. Why? Because he allowed the airplane to land without the passengers wearing seat belts. The passengers were all piled up in the back, none in a seat by themselves, and none with a seat belt. The pilot had placed the concern for the airplane over the concern for his passengers. We can always get another airplane; that's why we pay those large insurance premiums. But we cannot replace the people.

CHAPTER SIXTEEN

Pilot Personality

THE ACCIDENT NUMBERS have proven that a Killing Zone exists. Pilots are given the responsibility to make their own decisions, but some pilots are not ready to handle that responsibility. Like anything else, too much of a good thing is bad. The Killing Zone exists because freedom is too much of a good thing.

The accident numbers are only symptoms of the larger problem. The larger problem affects all pilots, even those that are not involved in accidents and therefore do not show up in accident statistics. Is it the problem deep seated in the personalities of those who fly?

There is a body of research now available that indicates that those who fly have different personality traits than everyone else. These personality traits are found in military and civilian pilots, professional and pleasure pilots, and both male and female pilots. When pilots are asked, "Why do you fly?" we give many answers. We say it's fun. We say we love it. We say its a getaway. But it really is deeper than that. It is usually something that is hard to put into words. If we could describe it and if we were honest with ourselves, we would say that we fly because flying offers the ingredients that our inner personality needs to thrive.

Eric Farmer, writing for the *International Journal of Aviation Safety*, says that a number of studies have supported the notion that a distinct pilot personality profile exists and can be identified in successful pilots. These studies have identified traits of the pilot personality. The research concludes that pilots are independent achievers who demonstrate competence and enjoy mastering complex tasks. The motivation to fly involves a need for prestige and for control. Those who love to fly desire excitement, power, speed, independence, thrills, and competition.

The people who hire pilots are very interested in what makes a competent pilot/employee. If there are personality traits indicative of a dangerous pilot, they would like to know this in advance so they can hire someone else. Writing in *Aviation, Space, and Environmental Medicine*, R. A. Alkov spoke of the "Personality Profile of Pilots." He points out that accidents are extremely costly to the industry not only in terms of loss of personnel and equipment, but also because the continued success of the industry, its very existence, depends on the public expectation of safe and reliable travel. The ability to identify the accident-prone pilot would be a valuable asset in the selection process. Factors associated with adventurousness and attitudes toward risk-taking were found to be significantly correlated with accident occurrence among pilots.

Today many airlines attempt to predict the future behavior of potential pilots with aptitude tests, personality tests, and psychological tests as a part of the interview process. Ronald Ferrara, a human factors researcher, has stated that, "In the [pilot] selection process it is important to differentiate between those who will be successful and those who will fail, as well as those who will remain active as professional pilots for a significant period of time. Selection of personnel involves the assessment of various factors in the individual's motivation to fly." This is why every pilot interview will include questions like, "What do you like about flying?" or "What got you interested in flying."

The safe pilot in general is mature, motivated to achieve, a well-integrated individual who has a positive self-image, is curious, active, and is able to cope with life's challenges. Researchers Joseph Novello and Zakhour Youssef conducted a study of both male and female pilots and identified some good and some bad characteristics of the pilot personality. The areas specifically identified were achievement, exhibitionism, dominance, the acceptance of change, adventure, deference, order, and endurance.

Achievement

Pilots are very goal oriented. A pilot sees a difficult task and enjoys working toward mastering the task. Working to achieve a piloting skill, which we perceive as difficult, feeds our need for personal reward. The pilot personality involves a strong drive to aim at and accomplish difficult tasks. The more difficult the task, the greater the reward. This is why we think very highly of ourselves when we pass a checkride, or fly into and out of a busy airport. But it can also be why we might try to "scud run" under a low overcast, or fly when we are not current.

Exhibition

Pilots not only want to achieve where others fail; they want recognition for the effort. The movies have portrayed the pilot as a "romantic hero." Real pilots know there is little romance in electrical systems, weight-and-balance forms, and performance charts, but we still don't mind the image. We want nonpilots to believe that we alone have what it takes to brave the heavens, to cheat death, and to reach out and touch the face of God. Pilots really like to talk about their own experiences, and certainly exaggeration is just part of the art. When we talk we like to make the point that what we're talking about is very dangerous and you really shouldn't try it yourself. For a person with the pilot personality, it is usually not enough just to please themselves; they must also put awe into the hearts of others. What endeavor other than flying can do that so readily? It is also this trait that pushes a well-meaning, otherwise safe pilot to buzz their friend's house.

Dominance

Pilots really like the control that flight gives them. There are very few situations in real life that give a person total and complete authority like the responsibility of being pilot in command. While on the ground, there are may people and circumstances that control me: my employer, my bills to pay, my family, my hectic schedule. But when I am at the controls of an airplane in flight, I am finally in command. I have power over the situation. The pilot personality needs to feel in control, needs to dominate the surroundings. Nowhere else can such power be obtained better than in an airplane. But being in control might actually blind a pilot to suggestions or options. A pilot in command could ignore a safer path.

Change

If you like the security of doing the same thing every day, in the same way, on the same schedule, then flying will not suit you. Pilots tend to enjoy doing new things. If you really want to frustrate pilots, make them do the same task repeatedly on an assembly line all day. The pilot personality gets bored quickly. Once again, piloting an airplane feeds the need for change. While in flight the situation is constantly in a state of change. Every flight, every landing, every instrument approach is different from the one before. A pilot sees each flight as being a completely new and different challenge that must be met and overcome. But there are some things that do take place routinely on any flight, like checklists and standard procedures. We are not just being forgetful if we miss a checklist item. It is in our personality to resist the repetition of doing the same checklist every time.

Adventure

When pilot groups are asked what they would rather do, (A) tour historic Europe, or B) sail around the world solo, they overwhelmingly like the idea of one person against the ocean. Pilots are more likely to seek adventure, to reach the unreachable star. Why climb Mount Everest? Because it is there! Pilots have this "because it's there" spirit. The thought of taking an airplane up against the elements, flying to distant places, and seeing things the average guy only dreams of, this is what drives pilots to the airport. But it can also drive a pilot past adventurer to thrill-seeker.

Flying offers the opportunity to satisfy our need for achievement, exhibitionism, dominance, the need to change, and adventure. We fly because it fits who we are. People who need goals set before them, or like to talk about personal adventures, or will push for their own point of view, those people can quench those cravings in an airplane. We fly because our inner, complex self is not happy with anything else. Flying has all the right ingredients.

But there is a dark side to all this. The pilot personality carries with it some potentially unhealthy baggage. Adventurism can be carried to a dangerous point. Control can reduce our teamwork. Exhibitionism can go beyond hangar flying to low-level aerobatics. Too much of a good thing is bad. And the pilot personality also has traits within it that are not good at any level. Pilots have the potential for faulty decision making that goes beyond poor judgment; it is actually part of their personality.

Deference

Pilots do not like to yield to another person's judgment or will. The pilot personality would rather take charge than be submissive. Pilots do not take criticism well; it goes against their grain. If you are in need of constructive criticism, they reason, then you are not in complete control. Because you want and need the control, pilots often reject the criticism or argue against it. This trait can be a real problem and a nightmare in crew resource management situations. If an instructor makes the suggestion that you shallow out your VOR intercept angle, your first reaction may be that he is just picking on you and that the angle you selected is satisfactory. Doing what the instructor has suggested would be an admission that someone else is in control.

Many of the accident examples used throughout this book involved brand-new pilots. They immediately went out and did something foolish soon after being taught to do otherwise. How could this happen? They got their license and for the first time they were no longer "under the thumb" of their flight instructor. They were now in complete control and they were going to do exactly what they wanted to without someone looking over their shoulder.

Order

Pilots as a group are not as interested in orderliness as the general population. They are not necessarily sloppy people, but putting things in the "proper" order is just not a very high priority for them. This trait could lead a pilot to overlook details on a preflight inspection or to only partially complete navigation plans. The decision to get into the air may take priority over organizing the cockpit for efficiency. Details become trivial. Items are either overlooked or rationalized as not important.

Endurance

Because the pilot personality likes change, it follows that anything that remains unchanged for long periods of time would be uncomfortable.

Unfortunately, the knowledge and skills of flying require the learner to show endurance over the long haul. Pilots must "stick to it" in order to gain any ground, but more often pilots will put aside a task when the first diversion comes along. The adventure-seeking traits that first bring the potential pilot to begin flight lessons may be overshadowed by the lack of endurance, and the would-be pilot never reaches the goal. There are many more student pilot certificates issued than private pilot certificates.

Do not think that just because you are a pilot that makes you an uncaring, sloppy, easily diverted, stubborn hardhead. But these traits, both good and bad, are part of what motivates us to fly. Taking all the traits together points out a dangerous inconsistency within the pilot personality. The same list of traits that drew a person to become a pilot in the first place can work against the pilot as well. Deep within the pilot personality are the reasons why we fly. We need to fly because flying satisfies and nurtures our personality. But also planted in our personality are the seeds of poor decision making. The same forces that bring us to the airport may also be our undoing. Put another way, a person who will never take a chance, never be too adventurous, and is perfectly meticulous would be a good pilot, but people with those traits seldom become pilots.

The reason there is a Killing Zone is part inexperience, but it is naive to place all the blame there. The Killing Zone also exists because of the pilot personality. The ultimate irony is that the same traits that create a pilot can kill a pilot.

Pilot Personality Self-Assessment Exercise

The following are 40 questions that have been used in identifying the pilot personality traits. This is an unscientific assessment, so don't base all your career plans on just this alone. But it is interesting to see how it comes out. Go through the exercise and be honest with yourself; be very honest.

1. If you were reincarnated and came back as an animal, which one would you want to be?
   a. Shark
   b. Bald Eagle
   c. House cat

2. What would be your choice of vacations?
   a. Sail around the world solo
   b. Wildlife safari in Africa
   c. Tour historic Europe
3. Who would you rather be?
   a. Astronaut "Hoot" Gibson
   b. Talk show host Oprah Winfrey
   c. Scientist Albert Einstein
4. Which challenge would you most likely want to accept?
   a. Relief pitcher coming in during the ninth inning of the seventh game of the World Series. Your team ahead 1-0. There are two outs and a man is on third.
   b. You are on the 18th green of the Master's Golf Tournament. You have a one-shot lead and face a tricky 10-foot putt for par.
   c. You are a contestant on "Who Wants to Be a Millionaire" and you face the final question for all the money.
5. When you do the laundry do you—
   a. Throw the clothes in a basket and fold them later
   b. Return to the dryer 30 minutes after the clothes have finished
   c. Neatly fold the clothes just as they finish drying
6. When you buy groceries for home do you—
   a. Go shopping when you are hungry and get just what you feel like having at the time
   b. Make a short list of items and pick them up on the way home
   c. Plan a week's worth of meals, buy only what is needed for the meals, and always stick to the plan
7. When it comes to being on time, people who know you would probably say—
   a. You are always late
   b. You are sometimes late, but you are busy and people understand
   c. They can set their watches by you
8. When it comes to "dumb" accidents you—
   a. Are considered accident prone
   b. Have your fair share
   c. Are pretty careful

9. What is the funniest?
   a. A pie-in-the-face gag
   b. A stand-up comedian
   c. A really clever political satire
10. You are—
    a. The life of the party
    b. A great networker
    c. A great listener
11. What is your attitude about people you know?
    a. Never completely trust anyone
    b. Only trust your closest friends
    c. You can't be happy without great friendships
12. It is 3 o'clock in the morning and you are stopped at a red light (that may be stuck). What would you do?
    a. Look both ways and go
    b. Wait for the light to turn green
    c. Call the police to report a broken traffic light
13. To help remember things you—
    a. Write reminders on scraps of paper and put the paper where you will see it
    b. Make a "to-do" list when you get really busy
    c. Complete items and then write them on a list just so you can cross them off
14. You like to play games—
    a. Alone
    b. With a partner
    c. As a part of a team
15. What do you want?
    a. Whatever I can get away with
    b. More than what my close friends have
    c. Nothing more than what is earned or owed
16. If you start a task you—
    a. Stop at the slightest excuse
    b. Work until something else comes up
    c. Cannot stop until you completely finish
17. When traveling by car you—
    a. Always drive no matter who else is in the car
    b. Prefer to drive

c. You don't care who drives

18. When you get lost while driving you—
   a. Use your superior sense of direction to get found again
   b. Consult a map
   c. Stop for directions

19. What best describes you?
   a. I am the best at whatever I do
   b. I am fun to be with
   c. I am set in my ways

20. When I drive I am always—
   a. Just over the speed limit, because I can drive better than most people
   b. Just over the speed limit, because the limit is just a little too conservative
   c. Drive the speed limit

21. The world would be much better off if—
   a. People were more like me
   b. Everyone would just mind their own business
   c. We could all be friends

22. When I see a hitchhiker I—
   a. Drive past and never give it a second thought
   b. Drive past but feel guilty that I didn't stop
   c. Give him a ride

23. You are flying with another pilot who has the same certificates and flight time as you. If the other pilot corrects you on something you will—
   a. Argue that what you did was better than what he is suggesting
   b. Listen but never invite her flying again
   c. Listen and try to learn something new

24. If you see someone make a mistake while flying you will—
   a. Point out the error and show them how it's done
   b. Quietly mention it to the person
   c. Say nothing if it does not affect the safety of the flight

25. When you feel strongly about something, you will—
   a. Push to have others see it your way
   b. Give your opinion if it comes up
   c. Keep it to yourself and say nothing

26. With friends you would rather—
    a. Tell others about your last flight to a busy airport
    b. Talk about airplanes you know about
    c. Listen while other pilots talk about their last solo cross country
27. Which would you rather do?
    a. Work on an assignment that gave you plenty of freedom to explore
    b. Work on an assignment in conjunction with committee members
    c. Work on an assignment that had strict guidelines and all you had to do was follow instructions
28. What situation would you enjoy most?
    a. A situation in which you had total control and complete authority
    b. A situation where group members contribute talents equally
    c. A situation where another person is in charge and you must take directions from that person
29. The job you would most likely enjoy would be one where—
    a. You do something different every day
    b. You do more thinking than physical labor
    c. You do mostly the same thing each day with no surprises
30. Filling out a complete navigation record, weight and balance form, and consulting several performance charts is—
    a. Good as a beginner but later unnecessary
    b. Good practice but not necessary on every flight
    c. Essential for safe flight every time
31. If the person who lives next to you complains that your music is too loud you—
    a. Tell the person that if they don't like it then they can move
    b. Invite them to join you
    c. Turn the music down
32. A pilot is best described as a—
    a. Thrill-seeker
    b. Romantic hero
    c. Just another person doing a job
33. A pilot is—
    a. A person with above-average talents who could excel in any field

    b. A person who wants more than an average life

    c. An average person who thinks flying is as good a way to make a living as anything else or the best way to beat the speed limit

34. A person should get into flying because—

    a. It's not what everybody does

    b. It is fun

    c. It can lead to big bucks

35. When a flight instructor gives you some constructive criticism about your flying techniques you—

    a. Feel that the instructor is just being picky

    b. Realize that your technique is not perfect but it's allowable

    c. Try to learn something from a person with more experience

36. When in flight training, your major objective is to—

    a. Know enough to pass the next checkride

    b. Impress the instructor

    c. Learn everything there is to learn

37. After you got your private pilot certificate, what would you have wanted to have happened?

    a. TV news in your hometown interviews you about your success

    b. Show your certificate to friends who are not pilots

    c. Nothing

38. What would give you the most pride?

    a. Getting a good grade, but doing so without your best effort

    b. Wasting an afternoon because you had nothing better to do

    c. Completing a difficult task, even when it takes 100% effort

39. You are most accused of forgetting—

    a. Where you put things

    b. Names, phone numbers, birthdays

    c. Faces

40. Why are you in flight training?

    a. Flying suits my personality

    b. I want a challenge to see if I can measure up

    c. I want a high-paying career

Now that you have completed the exercise, let's see how you did. Go back through the exercise. Every time you selected answer "a" give yourself one point. Answer "b" is two points, and answer "c" counts for three points. If you answered selection "b" on the first question (2 points), and selection a on the second question (1 point) you would have a total score of 3. Add the point value of all 40 questions and then divide by 40 to get a final average.

The closer your final average is to 1.0, the more that you have traits that have been identified in the pilot personality. Remember that this is not a foolproof exercise. If your average came out closer to 3.0, that does not mean you should give up flying; in fact you might even be a safer pilot. Keep in mind that not all of the traits of the pilot personality are good traits. Having a final average closer to 1.0 might confirm two important facts:

1. You have been attracted to flying because it helps feed who you are.
2. A person with your personality could get into trouble while flying or while making a decision about flying.

The dangerous thing about the pilot personality is that the very traits that make flying so fun, so exciting, so challenging, and that attract people to fly are the same traits that can make flying become dangerous. There is a line between being adventuresome and being a thrill-seeker and risk-taker. There is a line between being in command and being hardheaded. There is a line between seeking freedom and abusing that freedom. The pilot personality can be on either side of these lines. The goal is to mix in some discipline, some responsibility, and experience to make a safe pilot.

CHAPTER SEVENTEEN

Airmanship

I ONCE HEARD an air traffic controller describe his handling of an in-flight emergency. A pilot was forced to make an emergency landing at night when his engine threw a rod. Spotting an airport's green-and-white flashing beacon, he safely landed his airplane under the worst of conditions. The controller said, "That pilot showed a great deal of airmanship that night."

Airmanship is a combination of characteristics. It is hard to describe precisely; it is one of those things that you "know it when you see it." It involves at least one part pilot skill, one part judgment, one part knowledge, and one part experience. All of these parts are blended, and none can stand alone. Knowledge leads to judgment. Experience improves skills. Judgment keeps a pilot safe so the more experience may be gained. The more of each ingredient and the better the blend, the higher level of airmanship.

The pilots in the Killing Zone lack experience. When you have between 50 and 350 hours, you just have not seen very much yet or practiced enough yet. Figure 17.1 illustrates the trade-off between airmanship and accident potential. The accident numbers have shown that from 50 to 350 hours, a pilot's airmanship is low and the accident potential is high. But as airmanship is acquired, the accident potential is reduced. Airmanship prevents accidents.

Pilots with airmanship have the appropriate level of confidence. Novice pilots often are overconfident. Maybe they simply do not know enough to know that they should be worried. In many accident examples, inexperienced pilots took off and got into trouble when experienced

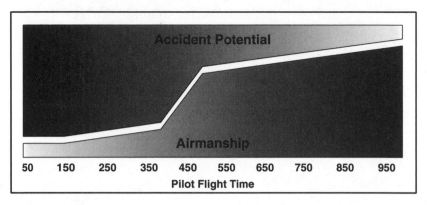

Fig. 17.1 The numbers don't lie. As airmanship is acquired, accident potential is reduced.

pilots stayed on the ground. Overconfidence and arrogance have been the real cause of many accidents among pilots in the Killing Zone. Arrogance and airmanship do not mix. If you have airmanship, you know better than to be arrogant. If you are arrogant, you have not learned enough to be an airman.

What is it about experience that is so valuable? And does it just automatically happen with time? The difference between an experienced pilot and a novice pilot is the quality of the decisions that they make. The decisions made by an experienced pilot are aided by the pilot's memory of past occurrences. If a person has experience, that means that they have seen more things. They have come across problems before and have seen what works and what does not work in different circumstances. If they ever come across that same problem again, they will already know what course of action to take because they have seen it before. Novice pilots have no firsthand background information to draw from, so they are reinventing the wheel as they go along. They don't know what solution has been proven, so they guess and hope for the best. Inexperienced pilots often guess incorrectly, with terrible outcomes.

I once was flying in IFR conditions with an instrument student and all at once our RPMs dropped from 2300 to about 1200. I pushed the throttle forward—no change. I pulled on carburetor heat—no help. We were on a downwind to an ILS approach, but we could no longer hold altitude. I told the controller that we had a problem and that we were turning to intercept the localizer right then. There was another airplane already on the approach, so the controller sent him around and got him out of the way. I intercepted the localizer inside the outer marker and intercepted the glide slope from above. We were able to hold the glide slope, and broke out of the clouds well above the decision height. We landed and I taxied over to the maintenance hangar. We shut down and my student, who had recently quit smoking, borrowed a cigarette from the A&P and lit up. After making sure the magnetos were off, the A&P mechanic pulled the propeller through by hand. The prop came to a point where it tightened up as one of the cylinders squeezing air on a compression stroke. Then the prop came through freely as the compression stroke of that cylinder was passed. He pulled the prop through again, and again it got hard to turn and then released. One more time, but this time the propeller did not get harder to turn. It passed through freely. The was no compression in whichever cylinder was on the compression stroke at that time. No compression means no power from that cylinder. We lost one of four cylinders, but almost half our power as the three working cylinders dragged the dead one along. We had swallowed a valve. The exhaust valve had broken off and fallen into the cylinder. Without the valve there was no way to seal the cylinder, so no compression was possible. What I learned was that pushing the throttle forward did not help and could have made it worse, since now a loose valve was being tossed around in the engine. Next time I'll know to reduce throttle. But I'll only know to do that because I have that flight as part of my experience now. If it ever happens again, I'm going to look very smart, but it's not because I'm smarter than anyone else; I'll just have been there before.

In an emergency there is little or no time to think through the problem. There is no time to make a list of pros and cons of your actions. If you take too much time to think, the window of opportunity to do anything may have passed. But if you have seen this situation before, you can remember what worked last time. Armed with this knowledge, you can begin corrective actions almost immediately. It will look to an observer that you have made a very complex decision in a spilt second, but in fact you are only applying a solution to a situation you recognize. The true definition of "experience" is when you recognize a wide range of potential problems and have ready-made solution matches for each one.

Does this mean that you must experience a problem before you can be experienced with the problem? I think not. I think that you can substitute knowledge for some level of experience. I think that motivation combined with good training and maturity can make a 300-hour pilot fly at the experience level of a 1000-hour pilot. If that is possible, then we have effectively removed the pilot from the Killing Zone. We have added additional parts of skill, knowledge, and judgment to replace a lack of experience in the mix. We have trained-out the next Killing Zone accident.

Many of the major U.S. airlines now know that it is possible to hire a pilot with 1500 hours who displays the airmanship of a 3000-hour pilot. I work with several major air carriers coordinating college student internships. The airlines know that they will be hiring many pilots in the years to come. They would like to hire young pilots because they get more return on their training investment if the pilot can fly for 35 years. Where do you get the best young piloting talent? The nation's aviation and aerospace universities is where they look, and internships is how they screen. They know that if they can get an eager-to-learn young pilot, they can turn them into dependable and safe employees/pilots. The airlines agree that a quality foundation in aviation education is worth thousands of hours of flight experience.

How does it work? How do you go about replacing a lack of experience with additional skill, knowledge, and judgment? First you play the "what-if" game. You ask yourself, or better yet, ask another pilot, "What if all of a sudden your RPMs dropped from 2300 to 1200?" "What if the temperature at your destination is 55 and the dew point is 50, and the sun is going down?" "What if your turn coordinator and attitude gyro do not agree?" You can think up a million of these. Discuss the what-if scenario. Get different points of view. Very often there will be more than one right answer and more than one course of action. If then someday in the air one of these what ifs comes true, you will be better prepared. You may not have seen it in flight, but you have "history" with the problem and you will not have to guess and hope for the best.

Next become active in ongoing flight training. Go to safety seminars, take aviation courses at a college, get into the wings program. Next, read. More information is available to pilots that at anytime in history from books, magazines, and the Internet. There is no need to write for FAA Advisory Circulars anymore; every one of them is now online. Every practical test standard is online, and every single question on every FAA knowledge test is online. Go to www.faa.gov and you will find it all. You can read the NASA reports of pilots who found themselves in dangerous situations and how they got out. You can read the National Transportation Safety Board accident investigations. If you haven't started increasing your knowledge using these methods, make it part of your five-year plan and get started.

As your flight time increases, you should attempt to increase the quality of your experience, not just your quantity. All flight hours are not the same. You receive a greater benefit toward airmanship when the challenge is greater. Chapter 13 illustrated that an hour seeking an advanced rating is worth more than an hour of personal flying. In general terms, an IFR flight is more valuable than a VFR flight. A night flight is worth more in experience gained than a daytime flight. A flight into a busy Class C airport is more helpful than a flight into an uncontrolled Class G airport. An hour in a complex airplane teaches more airmanship than an hour in a fixed-gear, fixed-pitch, airplane. Winter usually is more challenging than summer. A crosswind teaches more than a calm wind. You learn more on an instrument approach than on a visual approach. You get better at weather planning when the weather is marginal than when it is beautiful.

If you fly 100 hours back and forth between two uncontrolled airports during the day, you will receive some benefits. But you are selling yourself short. By raising the bar, by seeking out more challenges, your 100 hours will be worth more and you will acquire more airmanship. An hour of landings and proficiency work at night does not cost more than an hour in the day. Likewise, an hour spent working in a radar environment costs the same as an hour flying to a nonradar airport, but the value is more.

The toughest problem to solve in all of aviation is how to beat the Killing Zone. How do pilots without experience gain the experience without killing themselves in the process? The answer is to gain airmanship faster than flight hours.

CHAPTER EIGHTEEN

Air Safety and the Media

THE PRIMARY TARGET audience for this book has been pilots, especially pilots with limited experience who are flying through what was defined as a Killing Zone. But the book has statistical information that might also be helpful to members of the media who report on aviation safety.

As pilots we understand that the playing field is not level. Most car accidents are routine and therefore not news. An aircraft accident is rare and therefore newsworthy. But we do get tired of members of the media misrepresenting aviation. And perhaps we as pilots should know more of how the news media work. Let this chapter serve as a primer for even-handed and accurate media reporting on aviation.

Whenever there is an aircraft accident, members of the media and members of an investigation team automatically have conflicting motivation. The media need as many details as possible and they need them before a deadline. The problem is that in the early stages of any investigation, there are few details and no conclusions. The investigators know that what they turn up will affect the lives of many people. They do not want to create false starts. They do not want to imply something that may be overturned as new evidence is found. The course of the investigation can be misdirected and as a result the truth may remain hidden. So the investigators want to hold back at the very moment that the media want answers. In high-profile aircraft accidents, the media is planned for. Remember the great lengths the NTSB went to in accommodating the media during the search for John F. Kennedy, Jr.? But most aircraft accidents are low profile. These are usually investigated by the FAA field office inspectors and may be covered by local news. There is no formal media plan in these cases; it may be just one inspector doing everything. Members of the media should respect the investigators' need for time. We know that deadlines must be met, but in the past investigators have been portrayed by the media as uncooperative. Some have even implied that information was deliberately being withheld behind a screen of caution. When no answers are immediately forthcoming about what caused an airplane to crash, it's not because the investigators don't want the media to know; it's because they don't know yet. The investigation timetable and the media timetable are not the same. This is a fact of life that must be understood and respected.

If the investigation is not able to supply answers, do not stick a microphone in the face of just anyone. Every media outlet in every market should keep a list of aviation authorities in their area. This list should be compiled before a crisis occurs. Airport managers, ATC representatives, and FAA spokespersons should take the initiative to make media contacts in advance. Likewise, the media should seek out members of their community who might be helpful in an aviation emergency. The first on the list should be that area's FAA Safety Program Manager. Having lines of communication already established will provide a better flow of information when it's needed most. Without a predetermined game plan to receive accurate information, the media can look very foolish when they get a quote from someone who just happened by.

When the list of aviation authorities in an area is compiled, then work to develop a trust with the people on the list. I am quite often contacted by local media to comment on an accident, or a procedure, pertaining to an aviation topic. There are some members of the media that I will not work with and others that I am happy to help. The ones that I will not work with have betrayed my trust in the past.

When working with aviation experts, members of the media should remember that the motivation of the expert is to inform and better educate the public about aviation topics. They are not interested in getting involved in sensationalism. I did an interview for a local television station years ago when the recreational pilot certificate was first introduced. I made the point that even though the total flight hours was less for a recreational pilot, that safety would not be compromised because solo requirements were the same and the reduction in flight hours was linked to a reduction in privileges. The message I gave was that air safety was not in jeopardy. The station led off the newscast with the tease, "Joy riders will soon fill the skies!" I never worked with that station again. In the long run they hurt themselves because eventually that station was shut out by aviation authorities.

When working with an aviation authority, remember that some things really can't be properly explained in a sound bite. We understand there is just so much time in a newscast and just so many column inches in the newspaper, but if you chop and cut and splice what we really said into something that we didn't say, not only will you lose that expert for future stories, but it's just unfair and bad journalism.

It seems that bad news is always better for ratings than good news, but every effort should be made to report on aviation when there is no crisis. The Nashville Air Traffic Control Tower recently got a new, upgraded radar system. The control tower manager called several media outlets in town with the news that this new equipment, that would save lives, was up and running. The response? The media contacted were not interested in improved safety. But they said, "If it breaks down—give us a call!" Aviation affects the lives of every person living in this country, even if they never fly. The media earns no respect when they ignore that fact and show up only at the scene of an accident.

Very often media reports are so far off base that all future credibility is lost. Once a local television news crew thought they had a real scoop. They entered a small airport terminal with the cameras running. They deliberately made it look like this was some risky undercover mission as the camera peered around a corner behind the front desk. There on the corner of the counter was the Unicom Radio. The receptionist had stepped away from the desk as the camera rolled. The reporter said, "Our channel X cameras caught the air traffic control at the XXX airport unmanned!" The problem was that it was an uncontrolled airport. There are never any air traffic controllers at an uncontrolled airport. The station could have saved themselves a lot of embarrassment if they had just found out the real story instead of misrepresenting what was actually happening.

Another local television station thought they had uncovered a scandal involving taxpayer dollars and state-owned airplanes. The reporter discovered that a state commissioner had flown on a state-owned King Air to an outlying city on official business. The reporter compared the estimated cost of the King Air flight to the cost of an airline ticket and accused the state of reckless squandering of tax money. What the reporter failed to mention was that because the destination city was a small town, it was not directly served by any airline. To fly on one of the airlines, the commissioner would have had to have left the previous day, flown to another state, changed planes, flown to the nearest city, and then driven the rest of the way. He would have had to stay in a hotel the night before, eat, and then stay a second night because no flights departed after the meeting. In order to save the state air fare money, the state would have had to have paid for two nights, plus meals, and had a commissioner away from his office for three days instead of an afternoon. The King Air, if the truth had been told, saved the taxpayers of that state several hundred dollars. But the reporter was more interested in creating a scandal than reporting the truth.

Often when the media does report a positive story they sometimes get the emphasis wrong. Don't make a hero out of someone who overcame danger that they themselves created. A network television news magazine show once reported about the brave and heroic father who courageously led his son to safety after their airplane crashed in the mountains. But it was the father the night before that took off in bad weather, over rugged terrain, without oxygen, without a flight plan, and without enough fuel to reach the destination. If this father had placed his son in that much danger on the ground, the story would have been about child abuse.

When I hear media reports concerning aviation that expose the reporter's lack of knowledge, I often wonder if other professions are equally misrepresented. I won't know if a story on medicine is completely out of line. I imagine some doctor is watching the story and is as disappointed in the media as I am when it comes to aviation.

To maintain maximum credibility and to avoid looking foolish, here are some things to avoid:

- Don't call regional air carriers "puddle jumpers," making them sound like some old barnstormer. It makes the media look silly since these so-called "puddle jumpers" are today newer than those flown by major carriers.
- When "radar contact" is lost, that does not mean that an airplane has crashed. Radar uses a line-of-sight principle, so it cannot "see"

all the way to the ground in most locations. I fly at an airport that is 25 miles from the nearest radar installation. That means that each and every time I fly, I drop off the radar when I descend below about 600 feet above the ground. My radar contact is lost, but I am still flying safely, and routinely.

- Don't use a long lens beside parallel runways and then tell your audience that they are watching near midair collisions.
- Understand that the control tower controls only about 20% of air traffic. The rest is tucked away, sometimes not even at the airport, in radar rooms and air route traffic control centers.
- The terms stall and crash are often used together. But the meaning of the word "stall" is different to a pilot than it is to the general public. The public's association with the word stall is like when their car stalls. The general public defines a stall as an engine failure. But that is not what it means to a pilot. An aerodynamic stall has nothing to do with the engine. A glider can stall, and it doesn't even have an engine! So when the media report that an airplane stalled and crashed, the public thinks that airplane crashed because the engine quit. This is just not true. If an engine quits on any airplane, the airplane does not fall from the sky. The aerodynamics of lift are still in force. Once a pilot successfully glided a Boeing 737 to a safe off-airport landing near New Orleans. Both of his engines had failed, but the pilot flew the airplane down and all were unhurt.

When media representatives have a misunderstanding they often pass that misunderstanding on to their audiences. A recent headline reporting the fatal accident of an experimental airplane read, "Weight may have caused plane crash." The article goes on to report that, "Witnesses said the engine quit in the air, then the plane went into a flat spin and just fell straight out of the sky." There are several problems with this report. First of all, how did an overweight condition cause the engine to quit? The witness being quoted is giving an account of what he/she saw, but an engine failure does not cause a flat spin. When the general public reads an account like this without any further explanation, they start believing that if an engine ever quits, then automatically the airplane will crash. This is untrue and reporting it this way is misleading.

Our goal and the media's goal should be close to the same thing. We want the truth about aviation told, and the media want to report the truth. We have ground to work together. We want the true aviation story told so that the public will not needlessly worry about things that are not worth worrying about. We want more people to learn to fly, and we want more passengers to fly. The aviation industry is one of the world's safest forms of travel. When that message gets out, we all benefit. The media must get educated before they can educate. If you are a journalist, get some expert aviation advice before you report and then be fair with what you report. In the long run you will have people in aviation trust you and in turn you will have better, more accurate reporting in the future.

APPENDIX A

Total Fatal Accidents - Private & Student Pilots 1983 -2000 Figure 1.2	
0 to 49 flight hours	130
50 to 99 flight hours	259
100 to 149 flight hours	309
150 to 199 flight hours	233
200 to 249 flight hours	244
250 to 299 flight hours	184
300 to 349 flight hours	186
350 to 399 flight hours	109
400 to 449 flight hours	122
450 to 499 flight hours	98
500 to 549 flight hours	115
550 to 599 flight hours	64
600 to 649 flight hours	97
650 to 699 flight hours	55
700 to 749 flight hours	69
750 to 799 flight hours	57
800 to 849 flight hours	60
850 to 899 flight hours	27
900 to 949 flight hours	46
950 to 999 flight hours	37

Fig. A.1

Total Fatal Accidents - Weather Related 1964 - 1972 Figure 1.3	
0 to 100 flight hours	244
100 to 299 flight hours	476
300 to 599 flight hours	295
600 to 899 flight hours	154
900 to 1199 flight hours	109
1200 to 1499 flight hours	65
NTSB Special Study. Adopted August 28, 1974	

Fig. A.2

FAA Estimated General Aviation Flight Hours 1982 - 1999 Figure 2.1	
1982 flight hours	29,640,000
1983 flight hours	28,673,000
1984 flight hours	29,099,000
1985 flight hours	28,322,000
1986 flight hours	27,073,000
1987 flight hours	26,972,000
1988 flight hours	27,446,000
1989 flight hours	27,920,000
1990 flight hours	28,510,000
1991 flight hours	27,678,000
1992 flight hours	24,780,000
1993 flight hours	22,796,000
1994 flight hours	22,235,000
1995 flight hours	24,906,000
1996 flight hours	24,881,000
1997 flight hours	25,464,000
1998 flight hours	26,796,000
1999 flight hours	27,080,000

Fig. A.3

General Aviation Fatal Accidents/100,000 Flt Hrs 1982 - 1999 Figure 2.2	
1982	1.99
1983	1.94
1984	1.87
1985	1.75
1986	1.75
1987	1.65
1988	1.68
1989	1.53
1990	1.55
1991	1.56
1992	1.80
1993	1.74
1994	1.80
1995	1.64
1996	1.45
1997	1.39
1998	1.36
1999	1.26

Fig. A.4

All General Aviation Accidents/100,000 Flt Hrs 1982 - 1999 Figure 2.3	
1982	10.90
1983	10.73
1984	10.36
1985	9.66
1986	9.54
1987	9.25
1988	8.69
1989	7.98
1990	7.77
1991	7.85
1992	8.36
1993	8.94
1994	8.96
1995	8.23
1996	7.67
1997	7.28
1998	7.12
1999	7.05

Fig. A.5

Fatal Accidents of Pilots with 50 to 350 Flight Hrs 1983 - 1998 Figure 2.4	
1983	115
1984	119
1985	111
1986	105
1987	100
1988	101
1989	102
1990	94
1991	94
1992	73
1993	80
1994	67
1995	80
1996	63
1997	60
1998	56

Fig. A.6

Fatal Accidents - Weather as Broad Cause
Private & Student Pilots 1983 -2000
Figure 3.1

0 to 49 flight hours	19
50 to 99 flight hours	88
100 to 149 flight hours	120
150 to 199 flight hours	92
200 to 249 flight hours	84
250 to 299 flight hours	65
300 to 349 flight hours	64
350 to 399 flight hours	43
400 to 449 flight hours	38
450 to 499 flight hours	39
500 to 549 flight hours	43
550 to 599 flight hours	22
600 to 649 flight hours	41
650 to 699 flight hours	20
700 to 749 flight hours	20
750 to 799 flight hours	19
800 to 849 flight hours	22
850 to 899 flight hours	13
900 to 949 flight hours	16
950 to 999 flight hours	13

Fig. A.7

Fatal Accidents - Continued VFR into IMC Private & Student Pilots 1983 -2000 Figure 3.2	
0 to 49 flight hours	19
50 to 99 flight hours	59
100 to 149 flight hours	73
150 to 199 flight hours	60
200 to 249 flight hours	64
250 to 299 flight hours	49
300 to 349 flight hours	50
350 to 399 flight hours	32
400 to 449 flight hours	37
450 to 499 flight hours	31
500 to 549 flight hours	32
550 to 599 flight hours	17
600 to 649 flight hours	35
650 to 699 flight hours	17
700 to 749 flight hours	16
750 to 799 flight hours	17
800 to 849 flight hours	19
850 to 899 flight hours	11
900 to 949 flight hours	14
950 to 999 flight hours	8

Fig. A.8

All Maneuvering Accidents Private & Student Pilots 1983 -2000 Figure 4.1	
0 to 49 flight hours	45
50 to 99 flight hours	89
100 to 149 flight hours	93
150 to 199 flight hours	70
200 to 249 flight hours	71
250 to 299 flight hours	62
300 to 349 flight hours	60
350 to 399 flight hours	32
400 to 449 flight hours	44
450 to 499 flight hours	29
500 to 549 flight hours	47
550 to 599 flight hours	23
600 to 649 flight hours	34
650 to 699 flight hours	14
700 to 749 flight hours	28
750 to 799 flight hours	14
800 to 849 flight hours	14
850 to 899 flight hours	18
900 to 949 flight hours	19
950 to 999 flight hours	16

Fig. A.9

Fatal Takeoff Accidents - Private & Student Pilots
Single engine 1983 -2000
Figure 5.1

0 to 49 flight hours	10
50 to 99 flight hours	26
100 to 149 flight hours	31
150 to 199 flight hours	15
200 to 249 flight hours	23
250 to 299 flight hours	13
300 to 349 flight hours	22
350 to 399 flight hours	9
400 to 449 flight hours	10
450 to 499 flight hours	11
500 to 549 flight hours	11
550 to 599 flight hours	8
600 to 649 flight hours	9
650 to 699 flight hours	8
700 to 749 flight hours	5
750 to 799 flight hours	7
800 to 849 flight hours	4
850 to 899 flight hours	1
900 to 949 flight hours	4
950 to 999 flight hours	3

Fig. A.10

Fatal Landing Accidents Private & Student Pilots 1983 -2000 Figure 6.1	
0 to 49 flight hours	101
50 to 99 flight hours	207
100 to 149 flight hours	240
150 to 199 flight hours	181
200 to 249 flight hours	178
250 to 299 flight hours	128
300 to 349 flight hours	138
350 to 399 flight hours	85
400 to 449 flight hours	84
450 to 499 flight hours	69
500 to 549 flight hours	86
550 to 599 flight hours	45
600 to 649 flight hours	69
650 to 699 flight hours	45
700 to 749 flight hours	51
750 to 799 flight hours	41
800 to 849 flight hours	47
850 to 899 flight hours	21
900 to 949 flight hours	33
950 to 999 flight hours	26

Fig. A.11

Fatal Alcohol & Drug Accidents
Private & Student Pilots 1983 -2000
Figure 10.1

0 to 49 flight hours	1
50 to 99 flight hours	6
100 to 149 flight hours	9
150 to 199 flight hours	6
200 to 249 flight hours	7
250 to 299 flight hours	3
300 to 349 flight hours	2
350 to 399 flight hours	3
400 to 449 flight hours	0
450 to 499 flight hours	2
500 to 549 flight hours	1
550 to 599 flight hours	2
600 to 649 flight hours	2
650 to 699 flight hours	0
700 to 749 flight hours	1
750 to 799 flight hours	1
800 to 849 flight hours	0
850 to 899 flight hours	1
900 to 949 flight hours	4
950 to 999 flight hours	2

Fig. A.12

Fatal Night Accidents - Private & Student Pilots 1983 -2000 Figure 11.1	
0 to 49 flight hours	14
50 to 99 flight hours	51
100 to 149 flight hours	70
150 to 199 flight hours	52
200 to 249 flight hours	45
250 to 299 flight hours	35
300 to 349 flight hours	34
350 to 399 flight hours	24
400 to 449 flight hours	25
450 to 499 flight hours	20
500 to 549 flight hours	28
550 to 599 flight hours	9
600 to 649 flight hours	25
650 to 699 flight hours	12
700 to 749 flight hours	11
750 to 799 flight hours	10
800 to 849 flight hours	8
850 to 899 flight hours	4
900 to 949 flight hours	5
950 to 999 flight hours	6

Fig. A.13

Fatal Accidents 1983-2000		
	IFR rated Private Pilots Figure 13.1	Non IFR Private Pilots Figure 1.2
100 to 149 flight hours	4	309
150 to 199 flight hours	9	233
200 to 249 flight hours	18	244
250 to 299 flight hours	28	184
300 to 349 flight hours	34	186
350 to 399 flight hours	25	109
400 to 449 flight hours	29	122
450 to 499 flight hours	22	98
500 to 549 flight hours	37	115
550 to 599 flight hours	21	64
600 to 649 flight hours	24	97
650 to 699 flight hours	13	55
700 to 749 flight hours	20	69
750 to 799 flight hours	22	57
800 to 849 flight hours	18	60
850 to 899 flight hours	6	27
900 to 949 flight hours	15	46
950 to 999 flight hours	19	37

Fig. A.14

Nonfatal Accidents 1983 -2000 Minor and Severe Injury Figure 13.2		
	IFR rated Private Pilots	Non IFR Private Pilots
150 to 199 flight hours	13	388
200 to 249 flight hours	22	320
250 to 299 flight hours	36	237
300 to 349 flight hours	40	249
350 to 399 flight hours	41	202
400 to 449 flight hours	29	176
450 to 499 flight hours	22	157
500 to 549 flight hours	30	163
550 to 599 flight hours	27	119
600 to 649 flight hours	24	101
650 to 699 flight hours	26	86
700 to 749 flight hours	24	82
750 to 799 flight hours	26	73
800 to 849 flight hours	27	77
850 to 899 flight hours	18	66
900 to 949 flight hours	20	62
950 to 999 flight hours	19	58

Fig. A.15

Nonfatal Accidents 1983 -2000 Minor and Severe Injury Figure 13.3		
	Commercial Pilots	Non IFR Private Pilots
150 to 199 flight hours	1	237
200 to 249 flight hours	10	320
250 to 299 flight hours	23	237
300 to 349 flight hours	28	249
350 to 399 flight hours	32	202
400 to 449 flight hours	28	176
450 to 499 flight hours	30	157
500 to 549 flight hours	41	163
550 to 599 flight hours	35	119
600 to 649 flight hours	38	101
650 to 699 flight hours	18	86
700 to 749 flight hours	26	82
750 to 799 flight hours	20	73
800 to 849 flight hours	37	77
850 to 899 flight hours	30	66
900 to 949 flight hours	21	62
950 to 999 flight hours	22	58

Fig. A.16

Fatal Accidents - Student & Private Pilots Single engine high preformance & Multiengine Figure 15.1		
	Single engine high performance	Multiengine
0 to 49 flight hours	3	0
50 to 99 flight hours	26	4
100 to 149 flight hours	63	4
150 to 199 flight hours	55	4
200 to 249 flight hours	61	6
250 to 299 flight hours	60	6
300 to 349 flight hours	61	12
350 to 399 flight hours	41	9
400 to 449 flight hours	50	8
450 to 499 flight hours	45	12
500 to 549 flight hours	35	13
550 to 599 flight hours	18	11
600 to 649 flight hours	49	7
650 to 699 flight hours	26	4
700 to 749 flight hours	28	8
750 to 799 flight hours	28	7
800 to 849 flight hours	26	11
850 to 899 flight hours	17	7
900 to 949 flight hours	8	4
950 to 999 flight hours	16	2

Fig. A.17

Index

About the Author

Paul A. Craig, Ed.D., longtime pilot, flight instructor, and aviation educator and author, designed and conducted the extensive pilot study that uncovered the killing zone. Driven by a lifelong concern with the high accident rate among general aviation pilots, Dr. Craig ran the research project while earning his doctorate in education, with special emphasis on pilot decision-making and flight training. A Gold Seal Multiengine Flight Instructor and twice FAA District Flight Instructor of the Year, he has spoken widely to flight instructors and others on improving flight training and safety. He is the author of *Pilot in Command*; *Be a Better Pilot*; *Stalls & Spins*; *Multiengine Flying*, 2nd Edition; and *Light Airplane Navigation Essentials*, all from McGraw-Hill's renowned Practical Flying Series.